NEC3 and NEC4 Compared

NEC3 and NEC4 Compared

Robert Alan Gerrard
BSc (Hons), FRICS, FCIArb, FCInstCES

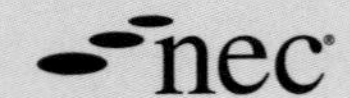

Published by ICE Publishing, One Great George Street, Westminster, London SW1P 3AA.

Full detail of ICE Publishing representatives and distributors can be found at:
www.icebookshop.com/bookshop_contact.asp

Other titles by ICE Publishing:

- *NEC4 Practical Solutions*
  R. Gerrard and S. Kings. ISBN: 978-0-7277-6199-6
- *Managing Reality, A practical guide to applying NEC4, Third Edition. 5-volume set.*
  B. Mitchell and B. Trebes. ISBN 978-0-7277-6195-8
- *NEC4: The Role of the Project Manager*
  B. Mitchell and B. Trebes. ISBN: 978-0-7277-6353-2

www.icebookshop.com
A catalogue record for this book is available from the British Library

ISBN 978-0-7277-6201-6

ICE Publishing is a division of Thomas Telford Ltd, a wholly-owned subsidiary of the Institution of Civil Engineers (ICE).

Commissioning Editor: Michael Fenton
Production Editor: Madhubanti Bhattacharyya
Market Development Executive: Elizabeth Hobson

Typeset by Manila Typesetting Company
Printed and bound in Great Britain by Bell and Bain, Glasgow

Contents

Introduction

NEC3 and NEC4 Compared is a clause-by-clause comparison of the changes between the *NEC3 Engineering and Construction Contract (ECC3)* and the *NEC4 Engineering and Construction Contract (ECC4)*. There has been considerable feedback since the launch of NEC3 and, to ensure a comprehensive review of all users' contributions, a number of working groups were set up to appraise comments and recommend actions. NEC then debated the principles of change and concluded that the time was right to produce updated versions of existing NEC3 documents together with new forms of contract to cover design build and operate contracts. These have now been published as the NEC4 suite of documents.

The aim of this book is to show those already using NEC3 the scale of changes between the NEC3 and NEC4 editions of the ECC, how these changes will affect them, and the implications of using NEC4. The principle behind the book is that the reader will be able to identify at first glance all information that is new and what has been deleted between the two editions.

The book presents the wording of ECC3 against that of ECC4 and provides a comprehensive guide to clauses in ECC3 that have been deleted, those that have been changed, and the clauses in ECC4 which are entirely new. It also shows where clauses have been re-ordered.

All information that has been deleted from ECC3 can be found on the left-hand side shaded in grey. Additional information or substantial changes that are new to ECC4 are underlined on the right-hand side. Where there is no corresponding ECC3 clause on the left-hand side, the ECC4 clause is entirely new. Where the position of a clause has changed the ECC3 clause has been re-ordered to align with the corresponding clause in the ECC4 contract. The book will therefore be easier to use if the reader refers back from clauses in ECC4 to the relevant ECC3 clause, rather than starting with the amended version of ECC3.

For completeness, the final version of ECC3 as published in April 2013 has been used for a direct comparison. There are additional explanatory notes under appropriate clauses in ECC4, giving details of more significant changes. These can be found in shaded boxes with an information symbol.

NEC4 comprises a suite of entirely consistent documents. The guide covers in detail the differences between the *NEC3 Engineering and Construction Contract* and the *NEC4 Engineering and Construction Contract*. However, principles of the changes it introduces have been applied across the main contracts in the suite of NEC4 contracts. So, anybody wishing to understand the new provisions applied across the suite of NEC4 contracts will benefit from this book.

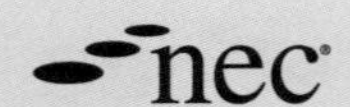

SCHEDULE OF OPTIONS

The strategy for choosing the form of contract starts with a decision between six main Options, one of which must be chosen.

Option A	Priced contract with activity schedule
Option B	Priced contract with bill of quantities
Option C	Target contract with activity schedule
Option D	Target contract with bill of quantities
Option E	Cost reimbursable contract
Option F	Management contract

One of the following dispute resolution Options must be selected to complete the chosen main Option.

Option W1	Dispute resolution procedure (used unless the United Kingdom Housing Grants, Construction and Regeneration Act 1996 applies).
Option W2	Dispute resolution procedure (used in the United Kingdom when the Housing Grants, Construction and Regeneration Act 1996 applies).

The following secondary Options should then be considered. It is not necessary to use any of them. Any combination other than those stated may be used.

Option X1	Price adjustment for inflation (used only with Options A, B, C and D)
Option X2	Changes in the law
Option X3	Multiple currencies (used only with Options A and B)
Option X4	Parent company guarantee
Option X5	Sectional Completion
Option X6	Bonus for early Completion
Option X7	Delay damages
Option X12	Partnering
Option X13	Performance bond
Option X14	Advanced payment to the *Contractor*
Option X15	Limitation of the *Contractor*'s liability for his design to reasonable skill and care
Option X16	Retention (not used with Option F)
Option X17	Low performance damages
Option X18	Limitation of liability
Option X20	Key Performance Indicators (not used with Option X12)

Schedule of Options

MAIN OPTIONS	The strategy for choosing the form of contract starts with a decision between six main Options, one of which must be chosen.
Option A	Priced contract with activity schedule
Option B	Priced contract with bill of quantities
Option C	Target contract with activity schedule
Option D	Target contract with bill of quantities
Option E	Cost reimbursable contract
Option F	Management contract
RESOLVING AND AVOIDING DISPUTES	One of the following procedures for resolving and avoiding disputes must be selected to complete the chosen main Option.
Option W1	Used where Adjudication is the method of dispute resolution and the United Kingdom Housing Grants, Construction and Regeneration Act 1996 does not apply
Option W2	Used where Adjudication is the method of dispute resolution and the United Kingdom Housing Grants, Construction and Regeneration Act 1996 applies
Option W3	Used where a Dispute Avoidance Board is the method of dispute resolution and the United Kingdom Housing Grants, Construction and Regeneration Act 1996 does not apply
SECONDARY OPTIONS	The following secondary Options should then be considered. It is not necessary to use any of them. Any combination other than those stated may be used.
Option X1	Price adjustment for inflation (used only with Options A, B, C and D)
Option X2	Changes in the law
Option X3	Multiple currencies (used only with Options A and B)
Option X4	Ultimate holding company guarantee
Option X5	Sectional Completion
Option X6	Bonus for early Completion
Option X7	Delay damages
Option X8	Undertakings to the *Client* or Others
Option X9	Transfer of rights Option
Option X10	Information Modelling
Option X11	Termination by the *Client*
Option X12	Multiparty collaboration (not used with X20)
Option X13	Performance bond
Option X14	Advanced payment to the *Contractor*
Option X15	The *Contractor's* design
Option X16	Retention (not used with Option F)
Option X17	Low performance damages
Option X18	Limitation of liability
Option X20	Key Performance Indicators (not used with Option X12)
Option X21	Whole Life Cost
Option X22	Early *Contractor* Involvement (used only with Options C and E)

The following Options dealing with national legislation should be included if required.

Option Y(UK)1	Project Bank Account
Option Y(UK)2	The Housing Grants, Construction and Regeneration Act 1996
Option Y(UK)3	The Contracts (Rights of Third Parties) Act 1999
Option Z	*Additional conditions of contract*
Note	Options X8 to X11 and X19 are not used.

The following Options dealing with national legislation should be included if required.

Option Y(UK)1	Project Bank Account
Option Y(UK)2	The Housing Grants, Construction and Regeneration Act 1996
Option Y(UK)3	The Contracts (Rights of Third Parties) Act 1999
Option Z	*Additional conditions of contract*
Note	Option X19 is not used

CORE CLAUSES

1 General

Actions **10**

10.1 The *Employer*, the *Contractor*, the *Project Manager* and the *Supervisor* shall act as stated in this contract and in a spirit of mutual trust and co-operation.

Identified and defined terms **11**

11.1 In these conditions of contract, terms identified in the Contract Data are in italics and defined terms have capital initials.

11.2 (1) The Accepted Programme is the programme identified in the Contract Data or is the latest programme accepted by the *Project Manager*. The latest programme accepted by the *Project Manager* supersedes previous Accepted Programmes.

(2) Completion is when the *Contractor* has

- done all the work which the Works Information states he is to do by the Completion Date and
- corrected notified Defects which would have prevented the *Employer* from using the *works* and Others from doing their work.

If the work which the *Contractor* is to do by the Completion Date is not stated in the Works Information, Completion is when the *Contractor* has done all the work necessary for the *Employer* to use the *works* and for Others to do their work.

(3) The Completion Date is the *completion date* unless later changed in accordance with this contract.

(4) The Contract Date is the date when this contract came into existence.

(5) A Defect is

- a part of the *works* which is not in accordance with the Works Information or
- a part of the *works* designed by the *Contractor* which is not in accordance with the applicable law or the *Contractor*'s design which the *Project Manager* has accepted.

(6) The Defects Certificate is either a list of Defects that the *Supervisor* has notified before the *defects date* which the *Contractor* has not corrected or, if there are no such Defects, a statement that there are none.

(14) The Risk Register is a register of the risks which are listed in the Contract Data and the risks which the *Project Manager* or the *Contractor* has notified as an early warning matter. It includes a description of the risk and a description of the actions which are to be taken to avoid or reduce the risk.

(7) Equipment is items provided by the *Contractor* and used by him to Provide the Works and which the Works Information does not require him to include in the *works*.

Core Clauses

1. GENERAL

Actions

10

10.1 The Parties, the *Project Manager* and the *Supervisor* shall act as stated in the contract.

10.2 The Parties, the *Project Manager* and the *Supervisor* act in a spirit of mutual trust and co- operation.

Clause 10.1 is now spilt into two clauses, the 'shall' being retained in only one of these clauses and only in relation to acting as stated in the contract. 

Identified and defined terms

11

11.1 In these *conditions of contract*, terms identified in the Contract Data are in italics and defined terms have capital initials.

11.2 (1) The Accepted Programme is the programme identified in the Contract Data or is the latest programme accepted by the *Project Manager*. The latest programme accepted by the *Project Manager* supersedes previous Accepted Programmes.

(2) Completion is when the *Contractor* has

- done all the work which the Scope states is to be done by the Completion Date and
- corrected notified Defects which would have prevented the *Client* from using the *works* or Others from doing their work.

If the work which the *Contractor* is to do by the Completion Date is not stated in the Scope, Completion is when the *Contractor* has done all the work necessary for the *Client* to use the *works* and for Others to do their work.

(3) The Completion Date is the *completion date* unless later changed in accordance with the contract.

(4) The Contract Date is the date when the contract came into existence.

(5) A Corrupt Act is

- the offering, promising, giving, accepting or soliciting of an advantage as an inducement for an action which is illegal, unethical or a breach of trust or
- abusing any entrusted power for private gain

in connection with this contract or any other contract with the *Client*. This includes any commission paid as an inducement which was not declared to the *Client* before the Contract Date.

This new defined term sets out what a Corrupt Act is. It is later referred to in clause 18, which forbids such acts, and again in clause 91.8, where the contract now provides for termination as a remedy. 

(6) A Defect is

- a part of the *works* which is not in accordance with the Scope or
- a part of the *works* designed by the *Contractor* which is not in accordance with the applicable law or the *Contractor's* design which the *Project Manager* has accepted.

(7) The Defects Certificate is either a list of Defects that the *Supervisor* has notified before the *defects date* which the *Contractor* has not corrected or, if there are no such Defects, a statement that there are none.

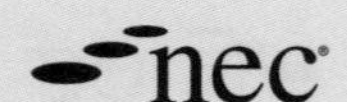

(8) The Fee is the sum of the amounts calculated by applying the *subcontracted fee percentage* to the Defined Cost of subcontracted work and the *direct fee percentage* to the Defined Cost of other work.

(9) A Key Date is the date by which work is to meet the Condition stated. The Key Date is the *key date* stated in the Contract Data and the Condition is the *condition* stated in the Contract Data unless later changed in accordance with this contract.

(10) Others are people or organisations who are not the *Employer*, the *Project Manager*, the *Supervisor*, the *Adjudicator*, the *Contractor* or any employee, Subcontractor or supplier of the *Contractor*.

(11) The Parties are the *Employer* and the *Contractor*.

(12) Plant and Materials are items intended to be included in the *works*.

(13) To Provide the Works means to do the work necessary to complete the *works* in accordance with this contract and all incidental work, services and actions which this contract requires.

(19) Works Information is information which either

- specifies and describes the *works* or
- states any constraints on how the *Contractor* Provides the Works

and is either

- in the documents which the Contract Data states it is in or
- in an instruction given in accordance with this contract.

(15) The Site is the area within the *boundaries of the site* and the volumes above and below it which are affected by work included in this contract.

(16) Site Information is information which

- describes the Site and its surroundings and
- is in the documents which the Contract Data states it is in.

(17) A Subcontractor is a person or organisation who has a contract with the *Contractor* to

- construct or install part of the *works,*
- provide a service necessary to Provide the Works or
- supply Plant and Materials which the person or organisation has wholly or partly designed specifically for the *works*.

(18) The Working Areas are those parts of the *working areas* which are

- necessary for Providing the Works and
- used only for work in this contract

unless later changed in accordance with this contract.

(8) The Early Warning Register is a register of matters which are

- listed in the Contract Data for inclusion and
- notified by the *Project Manager* or the *Contractor* as early warning matters.

It includes a description of the matter and the way in which the effects of the matter are to be avoided or reduced.

(9) Equipment is items provided and used by the *Contractor* to Provide the Works and which the Scope does not require the *Contractor* to include in the *works*.

(10) The Fee is the amount calculated by applying the *fee percentage* to the amount of Defined Cost.

(11) A Key Date is the date by which work is to meet the Condition stated. The Key Date is the *key date* stated in the Contract Data and the Condition is the *condition* stated in the Contract Data unless later changed in accordance with the contract.

(12) Others are people or organisations who are not the *Client,* the *Project Manager*, the *Supervisor*, the *Adjudicator* or a member of the Dispute Avoidance Board, the *Contractor* or any employee, Subcontractor or supplier of the *Contractor*.

(13) The Parties are the *Client* and the *Contractor*.

(14) Plant and Materials are items intended to be included in the *works*.

(15) To Provide the Works means to do the work necessary to complete the *works* in accordance with the contract and all incidental work, services and actions which the contract requires.

(16) Scope is information which

- specifies and describes the *works* or
- states any constraints on how the *Contractor* Provides the Works

and is either

- in the documents which the Contract Data states it is in or
- in an instruction given in accordance with the contract.

(17) The Site is the area within the *boundaries of the site* and the volumes above and below it which are affected by work included in the contract.

(18) Site Information is information which

- describes the Site and its surroundings and
- is in the documents which the Contract Data states it is in.

Some of the terms used across the NEC4 contracts have been standardised, so 'Works Information' has become 'Scope'. Note that '*Employer*' has become '*Client*' in another change aimed at standardisation.

CORE CLAUSES

MAIN OPTION CLAUSES

SECONDARY OPTION CLAUSES

COST COMPONENTS

Interpretation and the law **12**

12.1 In this contract, except where the context shows otherwise, words in the singular also mean in the plural and the other way round and words in the masculine also mean in the feminine and neuter.

12.2 This contract is governed by the *law of the contract*.

12.3 No change to this contract, unless provided for by the *conditions of contract*, has effect unless it has been agreed, confirmed in writing and signed by the Parties.

12.4 This contract is the entire agreement between the Parties.

Communications **13**

13.1 Each instruction, certificate, submission, proposal, record, acceptance, notification, reply and other communication which this contract requires is communicated in a form which can be read, copied and recorded. Writing is in the *language of this contract*.

13.2 A communication has effect when it is received at the last address notified by the recipient for receiving communications or, if none is notified, at the address of the recipient stated in the Contract Data.

13.3 If this contract requires the *Project Manager*, the *Supervisor* or the *Contractor* to reply to a communication, unless otherwise stated in this contract, he replies within the *period for reply*.

13.4 The *Project Manager* replies to a communication submitted or resubmitted to him by the *Contractor* for acceptance. If his reply is not acceptance, the *Project Manager* states his reasons and the *Contractor* resubmits the communication within the *period for reply* taking account of these reasons. A reason for withholding acceptance is that more information is needed in order to assess the *Contractor*'s submission fully.

13.5 The *Project Manager* may extend the *period for reply* to a communication if the *Project Manager* and the *Contractor* agree to the extension before the reply is due. The *Project Manager* notifies the *Contractor* of the extension which has been agreed.

13.6 The *Project Manager* issues his certificates to the *Employer* and the *Contractor*. The *Supervisor* issues his certificates to the *Project Manager* and the *Contractor*.

13.7 A notification which this contract requires is communicated separately from other communications.

13.8 The *Project Manager* may withhold acceptance of a submission by the *Contractor*. Withholding acceptance for a reason stated in this contract is not a compensation event.

(19) A Subcontractor is a person or organisation who has a contract with the *Contractor* to

- construct or install part of the *works*,
- design all or part of the *works*, except the design of Plant and Materials carried out by the supplier or
- provide a service in the Working Areas which is necessary to Provide the Works, except for the
  - hire of Equipment or
  - supply of people paid for by the *Contractor* according to the time they work.

The changes here are more about tightening up the previous definition of what is a Subcontractor, rather than trying to develop anything new.

(20) The Working Areas are the Site and those parts of the *working areas* which are

- necessary for Providing the Works and
- used only for work in the contract

unless later changed in accordance with the contract.

Interpretation and the law

12

12.1 In the contract, except where the context shows otherwise, words in the singular also mean in the plural and the other way round.

All masculine text has now been removed so the contract is on a gender-neutral basis, hence the amendments to clause 12.1.

12.2 The contract is governed by the *law of the contract*.

12.3 No change to the contract, unless provided for by these *conditions of contract*, has effect unless it has been agreed, confirmed in writing and signed by the Parties.

12.4 The contract is the entire agreement between the Parties.

Communications

13

13.1 Each communication which the contract requires is communicated in a form which can be read, copied and recorded. Writing is in the *language of the contract*.

13.2 If the Scope specifies the use of a communication system, a communication has effect when it is communicated through the communication system specified in the Scope.

If the Scope does not specify a communication system, a communication has effect when it is received at the last address notified by the recipient for receiving communications or, if none is notified, at the address of the recipient stated in the Contract Data.

Modern contracting often involves digital contract management and the changes to clause 13.2 provide for that.

13.3 If the contract requires the *Project Manager*, the *Supervisor* or the *Contractor* to reply to a communication, unless otherwise stated in these *conditions of contract*, they reply within the *period for reply*.

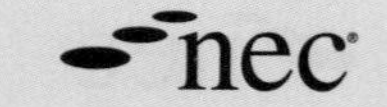

The *Project Manager* and the *Supervisor* **14**

14.1 The *Project Manager*'s or the *Supervisor*'s acceptance of a communication from the *Contractor* or of his work does not change the *Contractor*'s responsibility to Provide the Works or his liability for his design.

14.2 The *Project Manager* and the *Supervisor*, after notifying the *Contractor*, may delegate any of their actions and may cancel any delegation. A reference to an action of the *Project Manager* or the *Supervisor* in this contract includes an action by his delegate.

14.3 The *Project Manager* may give an instruction to the *Contractor* which changes the Works Information or a Key Date.

14.4 The *Employer* may replace the *Project Manager* or the *Supervisor* after he has notified the *Contractor* of the name of the replacement.

Early warning **16**

16.1 The *Contractor* and the *Project Manager* give an early warning by notifying the other as soon as either becomes aware of any matter which could

- increase the total of the Prices,
- delay Completion,
- delay meeting a Key Date or
- impair the performance of the *works* in use.

The *Contractor* may give an early warning by notifying the *Project Manager* of any other matter which could increase his total cost. The *Project Manager* enters early warning matters in the Risk Register. Early warning of a matter for which a compensation event has previously been notified is not required.

16.2 Either the *Project Manager* or the *Contractor* may instruct the other to attend a risk reduction meeting. Each may instruct other people to attend if the other agrees.

16.3 At a risk reduction meeting, those who attend co-operate in

- making and considering proposals for how the effect of the registered risks can be avoided or reduced,
- seeking solutions that will bring advantage to all those who will be affected,
- deciding on the actions which will be taken and who, in accordance with this contract, will take them and
- deciding which risks have now been avoided or have passed and can be removed from the Risk Register.

16.4 The *Project Manager* revises the Risk Register to record the decisions made at each risk reduction meeting and issues the revised Risk Register to the *Contractor*. If a decision needs a change to the Works Information, the *Project Manager* instructs the change at the same time as he issues the revised Risk Register.

13.4 The *Project Manager* replies to a communication submitted or resubmitted by the *Contractor* for acceptance. If the reply is not acceptance, the *Project Manager* states the reasons in sufficient detail to enable the *Contractor* to correct the matter. The *Contractor* resubmits the communication within the *period for reply* taking account of these reasons. A reason for withholding acceptance is that more information is needed in order to assess the *Contractor's* submission fully.

13.5 The *Project Manager* may extend the period for reply to a communication if the *Project Manager* and the *Contractor* agree to the extension before the reply is due. The *Project Manager* informs the *Contractor* of the extension which has been agreed.

Note that the Project Manager now 'informs' rather than 'notifies' the Contractor of the extension. This change in wording also occurs in other ECC4 clauses. 

13.6 The *Project Manager* issues certificates to the *Client* and the *Contractor*. The *Supervisor* issues certificates to the *Project Manager*, the *Client* and the *Contractor*.

13.7 A notification or certificate which the contract requires is communicated separately from other communications.

13.8 The *Project Manager* may withhold acceptance of a submission by the *Contractor*. Withholding acceptance for a reason stated in these *conditions of contract* is not a compensation event.

The *Project Manager* and the *Supervisor*

14

14.1 The *Project Manager's* or the *Supervisor's* acceptance of a communication from the *Contractor* or acceptance of the work does not change the *Contractor's* responsibility to Provide the Works or liability for its design.

14.2 The *Project Manager* and the *Supervisor*, after notifying the *Contractor*, may delegate any of their actions and may cancel any delegation. The notification contains the name of the delegate and details of the actions being delegated or any cancellation of delegation. A reference to an action of the *Project Manager* or the *Supervisor* in the contract includes an action by their delegate. The *Project Manager* and the *Supervisor* may take an action which they have delegated.

14.3 The *Project Manager* may give an instruction to the *Contractor* which changes the Scope or a Key Date.

14.4 The *Client* may replace the *Project Manager* or the *Supervisor* after notifying the *Contractor* of the name of the replacement.

Early warning

15

15.1 The *Contractor* and the *Project Manager* give an early warning by notifying the other as soon as either becomes aware of any matter which could

- increase the total of the Prices,
- delay Completion,
- delay meeting a Key Date or
- impair the performance of the *works* in use.

The *Project Manager* or the *Contractor* may give an early warning by notifying the other of any other matter which could increase the *Contractor's* total cost. The *Project Manager* enters early warning matters in the Early Warning Register. Early warning of a matter for which a compensation event has previously been notified is not required.

15.2 The *Project Manager* prepares a first Early Warning Register and issues it to the *Contractor* within one week of the *starting date*. The *Project Manager* instructs the *Contractor* to attend a first early warning meeting within two weeks of the *starting date*.

Later early warning meetings are held

- if either the *Project Manager* or *Contractor* instructs the other to attend an early warning meeting, and, in any case,

Adding to the Working Areas

15

15.1 The *Contractor* may submit a proposal for adding an area to the Working Areas to the *Project Manager* for acceptance. A reason for not accepting is that the proposed area is either not necessary for Providing the Works or used for work not in this contract.

- at no longer interval than the interval stated in the Contract Data until Completion of the whole of the *works*.

The *Project Manager* or *Contractor* may instruct other people to attend an early warning meeting if the other agrees.

A Subcontractor attends an early warning meeting if their attendance would assist in deciding the actions to be taken.

15.3 At an early warning meeting, those who attend co-operate in

- making and considering proposals for how the effects of each matter in the Early Warning Register can be avoided or reduced,
- seeking solutions that will bring advantage to all those who will be affected,
- deciding on the actions which will be taken and who, in accordance with the contract, will take them,
- deciding which matters can be removed from the Early Warning Register and
- reviewing actions recorded in the Early Warning Register and deciding if different actions need to be taken and who, in accordance with the contract, will take them.

15.4 The *Project Manager* revises the Early Warning Register to record the decisions made at each early warning meeting and issues the revised Early Warning Register to the *Contractor* within one week of the early warning meeting. If a decision needs a change to the Scope, the *Project Manager* instructs the change at the same time as the revised Early Warning Register is issued.

Quite a few changes have been made to the early warning process. The Risk Register has gone and is replaced with the Early Warning Register, a sensible move intended to prevent any confusion regarding the project's risk management process. There is also more prescription about when early warning meetings (rather than risk reduction meetings) are to be held. These still can be held instantly if the matter is urgent.

Contractor's proposals

16

16.1 The *Contractor* may propose to the *Project Manager* that the Scope provided by the *Client* is changed in order to reduce the amount the *Client* pays to the *Contractor* for the Providing the Works. The *Project Manager* consults with the *Client* and the *Contractor* about the change.

16.2 Within four weeks of the *Contractor* making the proposal the *Project Manager*

- accepts the *Contractor's* proposal and issues an instruction changing the Scope,
- informs the *Contractor* that the *Client* is considering the proposal and instructs the *Contractor* to submit a quotation for a proposed instruction to change the Scope or
- informs the *Contractor* that the proposal is not accepted.

The *Project Manager* may give any reason for not accepting the proposal.

Although not called as such, this brings into the core clauses a type of value engineering incentive mechanism. This was only previously available for Options C and D. Good communication in good time will be key to reaching a quick decision on whether to abandon the proposal or to work together to reach an agreement.

16.3 The *Contractor* may submit a proposal for adding an area to the Working Areas to the *Project Manager* for acceptance. A reason for not accepting is that the proposed area is

- not necessary for Providing the Works or
- used for work not in the contract.

CORE CLAUSES

MAIN OPTION CLAUSES

SECONDARY OPTION CLAUSES

COST COMPONENTS

Ambiguities and inconsistencies

17

17.1 The *Project Manager* or the *Contractor* notifies the other as soon as either becomes aware of an ambiguity or inconsistency in or between the documents which are part of this contract. The *Project Manager* gives an instruction resolving the ambiguity or inconsistency.

Illegal and impossible requirements

18

18.1 The *Contractor* notifies the *Project Manager* as soon as he considers that the Works Information requires him to do anything which is illegal or impossible. If the *Project Manager* agrees, he gives an instruction to change the Works Information appropriately.

Prevention

19

19.1 If an event occurs which

- stops the *Contractor* completing the *works* or
- stops the *Contractor* completing the *works* by the date shown on the Accepted Programme,

and which

- neither Party could prevent and
- an experienced contractor would have judged at the Contract Date to have such a small chance of occurring that it would have been unreasonable for him to have allowed for it,

the *Project Manager* gives an instruction to the *Contractor* stating how he is to deal with the event.

Requirements for instructions

17

17.1 The *Project Manager* or the *Contractor* notifies the other as soon as either becomes aware of an ambiguity or inconsistency in or between the documents which are part of the contract. The *Project Manager* states how the ambiguity or inconsistency should be resolved.

The changes in clause 17.1 recognise that the action may well be outside the powers of the *Project Manager* to resolve, such as a problem with the Option Z *additional conditions of contract*. The *Project Manager* can do certain things, such as instructing a change to the Scope, but other things might have to be changed by the Parties under clause 12.3.

17.2 The *Project Manager* or the *Contractor* notifies the other as soon as either becomes aware that the Scope includes an illegal or impossible requirement. If the Scope does include an illegal or impossible requirement, the *Project Manager* gives an instruction to change the Scope appropriately.

Clause 17.2 now allows the *Project Manager* to commence the process for dealing with an illegal or impossible requirement included in the Scope.

Corrupt Acts

18

18.1 The *Contractor* does not do a Corrupt Act.

18.2 The *Contractor* takes action to stop a Corrupt Act of a Subcontractor or supplier of which it is, or should be, aware.

18.3 The *Contractor* includes equivalent provisions to these in subcontracts and contracts for the supply of Plant and Materials and Equipment.

The subcontract requirements demand the *Contractor* takes positive action in such matters and that equivalent provisions are fed down the supply chain.

Prevention

19

19.1 If an event occurs which

- stops the *Contractor* completing the whole of the *works* or
- stops the *Contractor* completing the whole of the *works* by the date for planned Completion shown on the Accepted Programme,

and which

- neither Party could prevent and
- an experienced contractor would have judged at the Contract Date to have such a small chance of occurring that it would have been unreasonable to have allowed for it,

the *Project Manager* gives an instruction to the *Contractor* stating how the event is to be dealt with.

These slight changes confirm that such an event must affect the whole of the *works* and that it is the date for planned Completion as shown on the Accepted Programme that is used as one of the tests.

GLOSSARY OF NEW ECC4 TERMS	
Client	changed project role – formerly *Employer*
Corrupt Act	new defined term – defined in clause 11.2(5)
Dispute Avoidance Board	new entity, part of the revised dispute avoidance and resolution provisions
early warning meeting	changed contract term – formerly 'risk reduction meeting'
Early Warning Register	changed defined term – formerly 'Risk Register'
Scope	changed defined term – formerly 'Works Information'

CORE CLAUSES | MAIN OPTION CLAUSES | SECONDARY OPTION CLAUSES | COST COMPONENTS

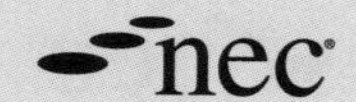

CORE CLAUSES

MAIN OPTION CLAUSES

SECONDARY OPTION CLAUSES

COST COMPONENTS

2 The *Contractor's* main responsibilities

Providing the Works

20

20.1 The *Contractor* Provides the Works in accordance with the Works Information.

The *Contractor's* design

21

21.1 The *Contractor* designs the parts of the *works* which the Works Information states he is to design.

21.2 The *Contractor* submits the particulars of his design as the Works Information requires to the *Project Manager* for acceptance. A reason for not accepting the *Contractor's* design is that it does not comply with either the Works Information or the applicable law.

The *Contractor* does not proceed with the relevant work until the *Project Manager* has accepted his design.

21.3 The *Contractor* may submit his design for acceptance in parts if the design of each part can be assessed fully.

Using the *Contractor's* design

22

22.1 The *Employer* may use and copy the *Contractor's* design for any purpose connected with construction, use, alteration or demolition of the *works* unless otherwise stated in the Works Information and for other purposes as stated in the Works Information.

Design of Equipment

23

23.1 The *Contractor* submits particulars of the design of an item of Equipment to the *Project Manager* for acceptance if the *Project Manager* instructs him to. A reason for not accepting is that the design of the item will not allow the *Contractor* to Provide the Works in accordance with

- the Works Information,
- the *Contractor's* design which the *Project Manager* has accepted or
- the applicable law.

People

24

24.1 The *Contractor* either employs each key person named to do the job stated in the Contract Data or employs a replacement person who has been accepted by the *Project Manager*. The *Contractor* submits the name, relevant qualifications and experience of a proposed replacement person to the *Project Manager* for acceptance. A reason for not accepting the person is that his relevant qualifications and experience are not as good as those of the person who is to be replaced.

24.2 The *Project Manager* may, having stated his reasons, instruct the *Contractor* to remove an employee. The *Contractor* then arranges that, after one day, the employee has no further connection with the work included in this contract.

Working with the *Employer* and Others

25

25.1 The *Contractor* co-operates with Others in obtaining and providing information which they need in connection with the *works*. He co-operates with Others and shares the Working Areas with them as stated in the Works Information.

2. THE *CONTRACTOR'S* MAIN RESPONSIBILITIES

Providing the Works

20

20.1 The *Contractor* Provides the Works in accordance with the Scope.

The *Contractor's* design

21

21.1 The *Contractor* designs the parts of the *works* which the Scope states the *Contractor* is to design.

21.2 The *Contractor* submits the particulars of its design as the Scope requires to the *Project Manager* for acceptance. A reason for not accepting the *Contractor's* design is that it does not comply with either the Scope or the applicable law.

The *Contractor* does not proceed with the relevant work until the *Project Manager* has accepted its design.

21.3 The *Contractor* may submit its design for acceptance in parts if the design of each part can be assessed fully.

Using the *Contractor's* design

22

22.1 The *Client* may use and copy the *Contractor's* design for any purpose connected with construction, use, alteration or demolition of the *works* unless otherwise stated in the Scope and for other purposes as stated in the contract. The *Contractor* obtains from a Subcontractor equivalent rights for the *Client* to use material prepared by the Subcontractor.

> Arguably the *Contractor* should have carried out the last sentence anyway as good practice, but this now makes the requirement quite clear.

Design of Equipment

23

23.1 The *Contractor* submits particulars of the design of an item of Equipment to the *Project Manager* for acceptance if the *Project Manager* instructs the *Contractor* to. A reason for not accepting is that the design of the item will not allow the *Contractor* to Provide the Works in accordance with

- the Scope,
- the *Contractor's* design which the *Project Manager* has accepted or
- the applicable law.

People

24

24.1 The *Contractor* either provides each *key person* named to do the job stated in the Contract Data or provides a replacement person who has been accepted by the *Project Manager*.

The *Contractor* submits the name, relevant qualifications and experience of a proposed replacement person to the *Project Manager* for acceptance. A reason for not accepting the person is that their relevant qualifications and experience are not as good as those of the person who is to be replaced.

24.2 The *Project Manager* may, having stated the reasons, instruct the *Contractor* to remove a person. The *Contractor* then arranges that, after one day, the person has no further connection with the work included in the contract.

Working with the *Client* and Others

25

25.1 The *Contractor* co-operates with Others, including in obtaining and providing information which they need in connection with the *works*. The *Contractor* shares the Working Areas with Others as stated in the Scope.

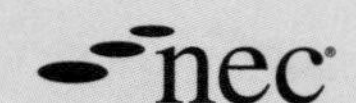

CORE CLAUSES
MAIN OPTION CLAUSES
SECONDARY OPTION CLAUSES
COST COMPONENTS

25.2 The *Employer* and the *Contractor* provide services and other things as stated in the Works Information. Any cost incurred by the *Employer* as a result of the *Contractor* not providing the services and other things which he is to provide is assessed by the *Project Manager* and paid by the *Contractor.*

25.3 If the *Project Manager* decides that the work does not meet the Condition stated for a Key Date by the date stated and, as a result, the *Employer* incurs additional cost either

- in carrying out work or
- by paying an additional amount to Others in carrying out work

on the same project, the additional cost which the *Employer* has paid or will incur is paid by the *Contractor*. The *Project Manager* assesses the additional cost within four weeks of the date when the Condition for the Key Date is met. The *Employer*'s right to recover the additional cost is his only right in these circumstances.

Subcontracting

26

26.1 If the *Contractor* subcontracts work, he is responsible for Providing the Works as if he had not subcontracted. This contract applies as if a Subcontractor's employees and equipment were the *Contractor*'s.

26.2 The *Contractor* submits the name of each proposed Subcontractor to the *Project Manager* for acceptance. A reason for not accepting the Subcontractor is that his appointment will not allow the *Contractor* to Provide the Works. The *Contractor* does not appoint a proposed Subcontractor until the *Project Manager* has accepted him.

26.3 The *Contractor* submits the proposed conditions of contract for each subcontract to the *Project Manager* for acceptance unless

- an NEC contract is proposed or
- the *Project Manager* has agreed that no submission is required.

The *Contractor* does not appoint a Subcontractor on the proposed subcontract conditions submitted until the *Project Manager* has accepted them. A reason for not accepting them is that

- they will not allow the *Contractor* to Provide the Works or
- they do not include a statement that the parties to the subcontract shall act in a spirit of mutual trust and co-operation.

25.2 The *Client* and the *Contractor* provide services and other things as stated in the Scope. Any cost incurred by the *Client* as a result of the *Contractor* not providing the services and other things which it is to provide is assessed by the *Project Manager* and paid by the *Contractor*.

25.3 If the *Project Manager* decides that the work does not meet the Condition stated for a Key Date by the date stated and, as a result, the *Client* incurs additional cost either

- in carrying out work or
- by paying an additional amount to Others in carrying out work

on the same project, the additional cost which the *Client* has paid or will incur is paid by the *Contractor*. The *Project Manager* assesses the additional cost within four weeks of the date when the Condition for the Key Date is met. The *Client's* right to recover the additional cost is its only right in these circumstances.

Subcontracting

26

26.1 If the *Contractor* subcontracts work, it is responsible for Providing the Works as if it had not subcontracted. The contract applies as if a Subcontractor's employees and equipment were the *Contractor's*.

26.2 The *Contractor* submits the name of each proposed Subcontractor to the *Project Manager* for acceptance. A reason for not accepting the Subcontractor is that the appointment will not allow the *Contractor* to Provide the Works. The *Contractor* does not appoint a proposed Subcontractor until the *Project Manager* has

- accepted the Subcontractor and, to the extent these *conditions of contract* requires,
- accepted the subcontract documents.

26.3 The *Contractor* submits the proposed subcontract documents, except any pricing information, for each subcontract to the *Project Manager* for acceptance unless

- the proposed subcontract is an NEC contract which has not been amended other than in accordance with the *additional conditions of contract* or
- the *Project Manager* has agreed that no submission is required.

A reason for not accepting the subcontract documents is that

- their use will not allow the *Contractor* to Provide the Works or
- they do not include a statement that the parties to the subcontract act in a spirit of mutual trust and co-operation.

Some general rewording is found in clause 26. Probably the most important change is that any proposed subcontract changes (secondary Option Z clauses) will be in keeping with the main contract. That is unless the *Project Manager* agrees no submission is required. This should help to keep contracts more closely aligned through the supply chain.

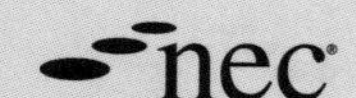

Other responsibilities

27

27.1 The *Contractor* obtains approval of his design from Others where necessary.

27.2 The *Contractor* provides access to work being done and to Plant and Materials being stored for this contract for

- the *Project Manager*,
- the *Supervisor* and
- Others notified to him by the *Project Manager*.

27.3 The *Contractor* obeys an instruction which is in accordance with this contract and is given to him by the *Project Manager* or the *Supervisor*.

27.4 The *Contractor* acts in accordance with the health and safety requirements stated in the Works Information.

Other responsibilities

27

27.1 The *Contractor* obtains approval of its design from Others where necessary.

27.2 The *Contractor* provides access to work being done and to Plant and Materials being stored for the contract for

- the *Project Manager*,
- the *Supervisor* and
- Others as named by the *Project Manager*.

27.3 The *Contractor* obeys an instruction which is in accordance with the contract and is given by the *Project Manager* or the *Supervisor*.

27.4 The *Contractor* acts in accordance with the health and safety requirements stated in the Scope.

Assignment

28

28.1 Either Party notifies the other Party if they intend to transfer the benefit of the contract or any rights under it. The *Client* does not transfer a benefit or rights if the party receiving the benefit or rights does not intend to act in a spirit of mutual trust and co-operation.

Disclosure

29

29.1 The Parties do not disclose information obtained in connection with the *works* except when necessary to carry out their duties under the contract.

29.2 The *Contractor* may publicise the *works* only with the *Client's* agreement.

Clauses 28 and 29 are both new. They are designed to reflect common practice by users, who prefer to allow for such provisions. These additions should help with standardisation and reduce the cost and burden of drafting secondary Option Z clauses commonly used in the industry.

GLOSSARY OF NEW ECC4 TERMS	
Client	changed project role – formerly *'Employer'*
key person	new identified term
proposed subcontract document	changed contract term – formerly 'proposed conditions of contract'
Scope	changed defined term – formerly 'Works Information'
Client	changed project role – formerly *'Employer'*
key person	new identified term

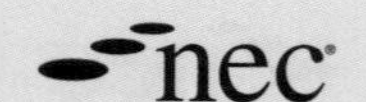

3 Time

Starting, Completion and Key Dates

30

30.1 The *Contractor* does not start work on the Site until the first *access date* and does the work so that Completion is on or before the Completion Date.

30.2 The *Project Manager* decides the date of Completion. The *Project Manager* certifies Completion within one week of Completion.

30.3 The *Contractor* does the work so that the Condition stated for each Key Date is met by the Key Date.

The programme

31

31.1 If a programme is not identified in the Contract Data, the *Contractor* submits a first programme to the *Project Manager* for acceptance within the period stated in the Contract Data.

31.2 The *Contractor* shows on each programme which he submits for acceptance

- the *starting date*, *access dates*, Key Dates and Completion Date,
- planned Completion,
- the order and timing of the operations which the *Contractor* plans to do in order to Provide the Works,
- the order and timing of the work of the *Employer* and Others as last agreed with them by the *Contractor* or, if not so agreed, as stated in the Works Information,
- the dates when the *Contractor* plans to meet each Condition stated for the Key Dates and to complete other work needed to allow the *Employer* and Others to do their work,
- provisions for
    - float,
    - time risk allowances,
    - health and safety requirements and
    - the procedures set out in this contract,
- the dates when, in order to Provide the Works in accordance with his programme, the *Contractor* will need
    - access to a part of the Site if later than its *access date*,
    - acceptances,
    - Plant and Materials and other things to be provided by the *Employer* and
    - information from Others,
- for each operation, a statement of how the *Contractor* plans to do the work identifying the principal Equipment and other resources which he plans to use and
- other information which the Works Information requires the *Contractor* to show on a programme submitted for acceptance.

3. TIME

Starting, Completion and Key Dates

30

30.1 The *Contractor* does not start work on the Site until the first *access date* and does the work so that Completion is on or before the Completion Date.

30.2 The *Project Manager* decides the date of Completion and certifies it within one week of the date.

30.3 The *Contractor* does the work so that the Condition stated for each Key Date is met by the Key Date.

The programme

31

31.1 If a programme is not identified in the Contract Data, the *Contractor* submits a first programme to the *Project Manager* for acceptance within the period stated in the Contract Data.

31.2 The *Contractor* shows on each programme submitted for acceptance

- the *starting date*, *access dates*, Key Dates and Completion Date,
- planned Completion,
- the order and timing of the operations which the *Contractor* plans to do in order to Provide the Works,
- the order and timing of the work of the *Client* and Others as last agreed with them by the *Contractor* or, if not so agreed, as stated in the Scope,
- the dates when the *Contractor* plans to meet each Condition stated for the Key Dates and to complete other work needed to allow the *Client* and Others to do their work,
- provisions for
  - float,
  - time risk allowances,
  - health and safety requirements and
  - the procedures set out in the contract,
- the dates when, in order to Provide the Works in accordance with the programme, the *Contractor* will need
  - access to a part of the Site if later than its *access date*,
  - acceptances,
  - Plant and Materials and other things to be provided by the *Client* and
  - information from Others,
- for each operation, a statement of how the *Contractor* plans to do the work identifying the principal Equipment and other resources which will be used and
- other information which the Scope requires the *Contractor* to show on a programme submitted for acceptance.

A programme issued for acceptance is in the form stated in Scope.

This additional sentence makes it clear that if the Scope does state the form the programme takes, e.g. a Gantt chart, then this must be provided. 

CORE CLAUSES

MAIN OPTION CLAUSES

SECONDARY OPTION CLAUSES

COST COMPONENTS

31.3 Within two weeks of the *Contractor* submitting a programme to him for acceptance, the *Project Manager* either accepts the programme or notifies the *Contractor* of his reasons for not accepting it. A reason for not accepting a programme is that

- the *Contractor*'s plans which it shows are not practicable,
- it does not show the information which this contract requires,
- it does not represent the *Contractor*'s plans realistically or
- it does not comply with the Works Information.

Revising the programme 32

32.1 The *Contractor* shows on each revised programme

- the actual progress achieved on each operation and its effect upon the timing of the remaining work,
- the effects of implemented compensation events,
- how the *Contractor* plans to deal with any delays and to correct notified Defects and
- any other changes which the *Contractor* proposes to make to the Accepted Programme.

The second bullet has been removed as this requirement is already covered in the first bullet. 

32.2 The *Contractor* submits a revised programme to the *Project Manager* for acceptance

- within the *period for reply* after the *Project Manager* has instructed him to,
- when the *Contractor* chooses to and, in any case,
- at no longer interval than the interval stated in the Contract Data from the *starting date* until Completion of the whole of the *works*.

Access to and use of the Site 33

33.1 The *Employer* allows access to and use of each part of the Site to the *Contractor* which is necessary for the work included in this contract. Access and use is allowed on or before the later of its *access date* and the date for access shown on the Accepted Programme.

Instructions to stop or not to start work 34

34.1 The *Project Manager* may instruct the *Contractor* to stop or not to start any work and may later instruct him that he may re-start or start it.

31.3 Within two weeks of the *Contractor* submitting a programme for acceptance, the *Project Manager* notifies the *Contractor* of the acceptance of the programme or the reasons for not accepting it. A reason for not accepting a programme is that

- the *Contractor's* plans which it shows are not practicable,
- it does not show the information which the contract requires,
- it does not represent the *Contractor's* plans realistically or
- it does not comply with the Scope.

If the *Project Manager* does not notify acceptance or non-acceptance within the time allowed, the *Contractor* may notify the *Project Manager* of that failure. If the failure continues for a further one week after the *Contractor's* notification, it is treated as acceptance by the *Project Manager* of the programme.

This new provision allows for the deemed acceptance in certain circumstances of a programme submitted for acceptance. This change is made to encourage the *Project Manager* to act in a timely fashion.

Revising the programme

32

32.1 The *Contractor* shows on each revised programme

- the actual progress achieved on each operation and its effect upon the timing of the remaining work,
- how the *Contractor* plans to deal with any delays and to correct notified Defects and
- any other changes which the *Contractor* proposes to make to the Accepted Programme.

32.2 The *Contractor* submits a revised programme to the *Project Manager* for acceptance

- within the *period for reply* after the *Project Manager* has instructed the *Contractor* to,
- when the *Contractor* chooses to and, in any case,
- at no longer interval than the interval stated in the Contract Data from the *starting date* until Completion of the whole of the *works*.

Access to and use of the Site

33

33.1 The *Client* allows access to and use of each part of the Site to the *Contractor* which is necessary for the work included in the contract. Access and use is allowed on or before the later of its *access date* and the date for access shown on the Accepted Programme.

Instructions to stop or not to start work

34

34.1 The *Project Manager* may instruct the *Contractor* to stop or not start any work. The *Project Manager* subsequently gives an instruction to the *Contractor* to

- re-start or start the work or
- remove the work from the Scope.

The changes to clause 34.1 include some general redrafting, but also now provide for a third scenario, which might be the removal of the work from the Scope. Arguably this could have been dealt with by two instructions (to restart or start, or remove), so this is a slight simplification.

CORE CLAUSES

Take over

35

35.1 The *Employer* need not take over the *works* before the Completion Date if it is stated in the Contract Data that he is not willing to do so. Otherwise the *Employer* takes over the *works* not later than two weeks after Completion.

35.2 The *Employer* may use any part of the *works* before Completion has been certified. If he does so, he takes over the part of the *works* when he begins to use it except if the use is

- for a reason stated in the Works Information or
- to suit the *Contractor*'s method of working.

35.3 The *Project Manager* certifies the date upon which the *Employer* takes over any part of the *works* and its extent within one week of the date.

Acceleration

36

36.1 The *Project Manager* may instruct the *Contractor* to submit a quotation for an acceleration to achieve Completion before the Completion Date. The *Project Manager* states changes to the Key Dates to be included in the quotation. A quotation for an acceleration comprises proposed changes to the Prices and a revised programme showing the earlier Completion Date and the changed Key Dates. The *Contractor* submits details of his assessment with each quotation.

36.2 The *Contractor* submits a quotation or gives his reasons for not doing so within the *period for reply*.

36.3 (Option A to D) When the *Project Manager* accepts a quotation for an acceleration, he changes the Prices, the Completion Date and the Key Dates accordingly and accepts the revised programme.

36.4 (Option E and F) When the *Project Manager* accepts a quotation for an acceleration, he changes the Completion Date, the Key Dates and the forecast of the total Defined Cost of the whole of the *works* accordingly and accepts the revised programme.

Take over **35**

35.1 The *Client* need not take over the *works* before the Completion Date if the Contract Data states it is not willing to do so. Otherwise the *Client* takes over the *works* not later than two weeks after Completion.

35.2 The *Client* may use any part of the *works* before Completion has been certified. The *Client* takes over the part of the *works* when it begins to use it except if the use is

- for a reason stated in the Scope or
- to suit the *Contractor's* method of working.

35.3 The *Project Manager* certifies the date upon which the *Client* takes over any part of the *works* and its extent within one week of the date.

Acceleration **36**

36.1 The *Contractor* and the *Project Manager* may propose to the other an acceleration to achieve Completion before the Completion Date. If the *Project Manager* and *Contractor* are prepared to consider the proposed change, the *Project Manager* instructs the *Contractor* to provide a quotation. The instruction states changes to the Key Dates to be included in the quotation. The *Contractor* provides a quotation within three weeks of the instruction to do so. The *Project Manager* replies to the quotation within three weeks. The reply is

- a notification that the quotation is accepted or
- a notification that the quotation is not accepted and that the Completion Dates and Key Dates are not changed.

36.2 A quotation for acceleration comprises proposed changes to the Prices and a revised programme showing the earlier Completion Date and the changed Key Dates. The *Contractor* submits details of the assessment with each quotation.

36.3 When a quotation for an acceleration is accepted, the *Project Manager* changes the Prices, the Completion Date and the Key Dates accordingly and accepts the revised programme.

Clause 36 has been re-drafted and all provisions are now found in the core clauses. The clause also starts off the process with a general agreement by the *Project Manager* and the *Contractor* to consider the change, rather than by the *Project Manager* instructing a quote to be provided only to find the *Contractor* does not want to accelerate.

GLOSSARY OF NEW ECC4 TERMS	
Client	changed project role – formerly '*Employer*'
Scope	changed defined term – formerly 'Works Information'

4 Testing and Defects

Tests and inspections

40

40.1 This clause only applies to tests and inspections required by the Works Information or the applicable law.

40.2 The *Contractor* and the *Employer* provide materials, facilities and samples for tests and inspections as stated in the Works Information.

40.3 The *Contractor* and the *Supervisor* each notifies the other of each of his tests and inspections before it starts and afterwards notifies the other of its results. The *Contractor* notifies the *Supervisor* in time for a test or inspection to be arranged and done before doing work which would obstruct the test or inspection. The *Supervisor* may watch any test done by the *Contractor*.

40.4 If a test or inspection shows that any work has a Defect, the *Contractor* corrects the Defect and the test or inspection is repeated.

40.5 The *Supervisor* does his tests and inspections without causing unnecessary delay to the work or to a payment which is conditional upon a test or inspection being successful. A payment which is conditional upon a *Supervisor*'s test or inspection being successful becomes due at the later of the *defects date* and the end of the last *defect correction period* if

- the *Supervisor* has not done the test or inspection and
- the delay to the test or inspection is not the *Contractor*'s fault.

40.6 The *Project Manager* assesses the cost incurred by the *Employer* in repeating a test or inspection after a Defect is found. The *Contractor* pays the amount assessed.

Testing and inspection before delivery

41

41.1 The *Contractor* does not bring to the Working Areas those Plant and Materials which the Works Information states are to be tested or inspected before delivery until the *Supervisor* has notified the *Contractor* that they have passed the test or inspection.

4. QUALITY MANAGEMENT

This section has a new title and a new clause 40 dealing with the *Contractor's* quality management system. A similar provision existed in other NEC3 contracts, but this is now applied consistently across the NEC4 main contracts. 

Quality management system

40

40.1 The *Contractor* operates a quality management system which complies with the requirements stated in the Scope.

40.2 Within the period stated in the Contract Data, the *Contractor* provides the *Project Manager* with a quality policy statement and a quality plan for acceptance. A reason for not accepting a quality policy statement or quality plan is that it does not allow the *Contractor* to Provide the Works.

If any changes are made to the quality plan, the *Contractor* provides the *Project Manager* with the changed quality plan for acceptance.

40.3 The *Project Manager* may instruct the *Contractor* to correct a failure to comply with the quality plan. This instruction is not a compensation event.

Tests and inspections

41

41.1 This clause only applies to tests and inspections required by the Scope or the applicable law.

41.2 The *Contractor* and the *Client* provide materials, facilities and samples for tests and inspections as stated in the Scope.

41.3 The *Contractor* and the *Supervisor* informs the other of each of their tests and inspections before the test or inspection starts and afterwards informs the other of the results. The *Contractor* informs the *Supervisor* in time for a test or inspection to be arranged and done before doing work which would obstruct the test or inspection. The *Supervisor* may watch any test done by the *Contractor*.

Note that the *Contractor* and the *Supervisor* now 'informs' rather than 'notifies' the other of its actions under this clause. 

41.4 If a test or inspection shows that any work has a Defect, the *Contractor* corrects the Defect and the test or inspection is repeated.

41.5 The *Supervisor* does tests and inspections without causing unnecessary delay to the work or to a payment which is conditional upon a test or inspection being successful. A payment which is conditional upon a *Supervisor's* test or inspection being successful becomes due at the later of the *defects date* and the end of the last *defect correction period* if

- the *Supervisor* has not done the test or inspection and
- the delay to the test or inspection is not the *Contractor's* fault.

41.6 The *Project Manager* assesses the cost incurred by the *Client* in repeating a test or inspection after a Defect is found. The *Contractor* pays the amount assessed.

Testing and inspection before delivery

42

42.1 The *Contractor* does not bring to the Working Areas those Plant and Materials which the Scope states are to be tested or inspected before delivery until the *Supervisor* has notified the *Contractor* that they have passed the test or inspection.

Searching for and notifying Defects

42

42.1 Until the *defects date,* the *Supervisor* may instruct the *Contractor* to search for a Defect. He gives his reason for the search with his instruction. Searching may include

- uncovering, dismantling, re-covering and re-erecting work,
- providing facilities, materials and samples for tests and inspections done by the *Supervisor* and
- doing tests and inspections which the Works Information does not require.

42.2 Until the *defects date*, the *Supervisor* notifies the *Contractor* of each Defect as soon as he finds it and the *Contractor* notifies the *Supervisor* of each Defect as soon as he finds it.

Correcting Defects

43

43.1 The *Contractor* corrects a Defect whether or not the *Supervisor* notifies him of it.

43.2 The *Contractor* corrects a notified Defect before the end of the *defect correction period*. The *defect correction period* begins at Completion for Defects notified before Completion and when the Defect is notified for other Defects.

43.3 The *Supervisor* issues the Defects Certificate at the later of the *defects date* and the end of the last *defect correction period*. The *Employer's* rights in respect of a Defect which the *Supervisor* has not found or notified are not affected by the issue of the Defects Certificate.

43.4 The *Project Manager* arranges for the Employer to allow the *Contractor* access to and use of a part of the *works* which he has taken over if they are needed for correcting a Defect. In this case the *defect correction period* begins when the necessary access and use have been provided.

Accepting Defects

44

44.1 The *Contractor* and the *Project Manager* may each propose to the other that the Works Information should be changed so that a Defect does not have to be corrected.

44.2 If the *Contractor* and the *Project Manager* are prepared to consider the change, the *Contractor* submits a quotation for reduced Prices or an earlier Completion Date or both to the *Project Manager* for acceptance. If the *Project Manager* accepts the quotation, he gives an instruction to change the Works Information, the Prices and the Completion Date accordingly.

Searching for and notifying Defects **43**

43.1 Until the *defects date,* the *Supervisor* may instruct the *Contractor* to search for a Defect. The *Supervisor* gives reasons for the search with the instruction. Searching may include

- uncovering, dismantling, re-covering and re-erecting work,
- providing facilities, materials and samples for tests and inspections done by the *Supervisor* and
- doing tests and inspections which the Scope does not require.

43.2 Until the *defects date* the *Supervisor* and the *Contractor* notifies the other as soon as they become aware of a Defect.

This change is an example of simplification and fine tuning of the drafting. 

Correcting Defects **44**

44.1 The *Contractor* corrects a Defect whether or not the *Supervisor* has notified it.

44.2 The *Contractor* corrects a notified Defect before the end of the *defect correction period.* The *defect correction period* begins at Completion for Defects notified before Completion and when the Defect is notified for other Defects.

44.3 The *Supervisor* issues the Defects Certificate at the *defects date* if there are no notified Defects, or otherwise at the earlier of

- the end of the last *defect correction period* and
- the date when all notified Defects have been corrected.

This minor change provides for when there are notified Defects that have not been corrected at the *defects date* but that are corrected before the end of the last *defect correction period.* The Defects Certificate is issued at this point, rather than at the end of the last *defect correction period*, which could be some time later. 

The *Client's* rights in respect of a Defect which the *Supervisor* has not found or notified are not affected by the issue of the Defects Certificate.

44.4 The *Project Manager* arranges for the *Client* to allow the *Contractor* access to and use of a part of the *works* which has been taken over if it is needed for correcting a Defect. In this case the *defect correction period* begins when the necessary access and use have been provided.

Accepting Defects **45**

45.1 The *Contractor* and the *Project Manager* may propose to the other that the Scope should be changed so that a Defect does not have to be corrected.

45.2 If the *Contractor* and the *Project Manager* are prepared to consider the change, the *Contractor* submits a quotation for reduced Prices or an earlier Completion Date or both to the *Project Manager* for acceptance. If the quotation is accepted, the *Project Manager* changes the Scope, the Prices, the Completion Date and the Key Dates accordingly and accepts the revised programme.

This slight change also reduces the amount of communication necessary in this instance – the acceptance does the job of changing the Scope etc., whereas previously there was an acceptance followed by an instruction. 

Uncorrected Defects **45**

45.1 If the *Contractor* is given access in order to correct a notified Defect but he has not corrected it within its *defect correction period*, the *Project Manager* assesses the cost to the *Employer* of having the Defect corrected by other people and the *Contractor* pays this amount. The Works Information is treated as having been changed to accept the Defect.

45.2 If the *Contractor* is not given access in order to correct a notified Defect before the *defects date*, the *Project Manager* assesses the cost to the *Contractor* of correcting the Defect and the *Contractor* pays this amount. The Works Information is treated as having been changed to accept the Defect.

Uncorrected Defects

46

46.1 If the *Contractor* is given access in order to correct a notified Defect but the Defect is not corrected within its *defect correction period*, the *Project Manager* assesses the cost to the *Client* of having the Defect corrected by other people and the *Contractor* pays this amount. The Scope is treated as having been changed to accept the Defect.

46.2 If the *Contractor* is not given access in order to correct a notified Defect before the *defects date*, the *Project Manager* assesses the cost to the *Contractor* of correcting the Defect and the *Contractor* pays this amount. The Scope is treated as having been changed to accept the Defect.

GLOSSARY OF NEW ECC4 TERMS	
Client	changed project role – formerly '*Employer*'
Quality management	changed section title – formerly 'Testing and Defects'
quality management system	changed clause title – formerly 'Tests and inspections'
Scope	changed defined term – formerly 'Works Information'

5 Payment

Assessing the amount due

50

50.1 The *Project Manager* assesses the amount due at each assessment date. The first assessment date is decided by the *Project Manager* to suit the procedures of the Parties and is not later than the *assessment interval* after the *starting date*. Later assessment dates occur

- at the end of each *assessment interval* until four weeks after the *Supervisor* issues the Defects Certificate and
- at Completion of the whole of the *works*.

50.4 In assessing the amount due, the *Project Manager* considers any application for payment the *Contractor* has submitted on or before the assessment date. The *Project Manager* gives the *Contractor* details of how the amount due has been assessed.

50.2 The amount due is

- the Price for Work Done to Date,
- plus other amounts to be paid to the *Contractor*,
- less amounts to be paid by or retained from the *Contractor*.

Any tax which the law requires the *Employer* to pay to the *Contractor* is included in the amount due.

50.3 If no programme is identified in the Contract Data, one quarter of the Price for Work Done to Date is retained in assessments of the amount due until the *Contractor* has submitted a first programme to the *Project Manager* for acceptance showing the information which this contract requires.

50.5 The *Project Manager* corrects any wrongly assessed amount due in a later payment certificate.

Payment

51

51.1 The *Project Manager* certifies a payment within one week of each assessment date. The first payment is the amount due. Other payments are the change in the amount due since the last payment certificate. A payment is made by the *Contractor* to the *Employer* if the change reduces the amount due. Other payments are made by the *Employer* to the *Contractor*. Payments are in the *currency of this contract* unless otherwise stated in this contract.

5. PAYMENT

Assessing the amount due

50

50.1 The *Project Manager* assesses the amount due at each assessment date. The first assessment date is decided by the *Project Manager* to suit the procedures of the Parties and is not later than the *assessment interval* after the *starting date*. Later assessment dates occur at the end of each *assessment interval* until

- the *Supervisor* issues the Defects Certificate or
- the *Project Manager* issues a termination certificate.

50.2 The *Contractor* submits an application for payment to the *Project Manager* before each assessment date setting out the amount the *Contractor* considers is due at the assessment date. The *Contractor's* application for payment includes details of how the amount has been assessed and is in the form stated in the Scope.

The *Contractor* is now obliged to submit an application for payment before each assessment date, not on the date. The previous omission of this obligation would have been a surprise to many users. 

In assessing the amount due, the *Project Manager* considers an application for payment submitted by the *Contractor* before the assessment date.

50.3 If the *Contractor* submits an application for payment before the assessment date, the amount due at the assessment date is

- the Price for Work Done to Date,
- plus other amounts to be paid to the *Contractor*,
- less amounts to be paid by or retained from the *Contractor*.

50.4 If the *Contractor* does not submit an application for payment before the assessment date, the amount due at the assessment date is the lesser of

- the amount the *Project Manager* assesses as due at the assessment date, assessed as though the *Contractor* had submitted an application before the assessment date, and
- the amount due at the previous assessment date.

Should the *Contractor* not submit an application for payment, it is very unlikely that anything will be paid for that particular assessment. However, it is unlikely that any *Contractor* will not submit an application, this being a key internal requirement. 

50.5 If no programme is identified in the Contract Data, one quarter of the Price for Work Done to Date is retained in assessments of the amount due until the *Contractor* has submitted a first programme to the *Project Manager* for acceptance showing the information which the contract requires.

50.6 The *Project Manager* corrects any incorrectly assessed amount due in a later payment certificate.

Payment

51

51.1 The *Project Manager* certifies a payment within one week of each assessment date. The *Project Manager's* certificate includes details of how the amount due has been assessed. The first payment is the amount due. Other payments are the change in the amount due since the previous assessment. A payment is made by the *Contractor* to the *Client* if the change reduces the amount due. Other payments are made by the *Client* to the *Contractor*. Payments are in the *currency of the contract* unless otherwise stated in the contract.

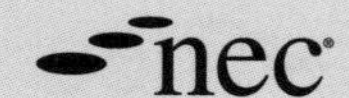

51.2 Each certified payment is made within three weeks of the assessment date or, if a different period is stated in the Contract Data, within the period stated. If a certified payment is late, or if a payment is late because the *Project Manager* does not issue a certificate which he should issue, interest is paid on the late payment. Interest is assessed from the date by which the late payment should have been made until the date when the late payment is made, and is included in the first assessment after the late payment is made.

51.3 If an amount due is corrected in a later certificate either

- by the *Project Manager* in relation to a mistake or a compensation event or
- following a decision of the *Adjudicator* or the *tribunal*,

interest on the correcting amount is paid. Interest is assessed from the date when the incorrect amount was certified until the date when the correcting amount is certified and is included in the assessment which includes the correcting amount.

51.4 Interest is calculated on a daily basis at the *interest rate* and is compounded annually.

Defined Cost

52

52.1 All the *Contractor*'s costs which are not included in the Defined Cost are treated as included in the Fee. Defined Cost includes only amounts calculated using rates and percentages stated in the Contract Data and other amounts at open market or competitively tendered prices with deductions for all discounts, rebates and taxes which can be recovered.

51.2 Each certified payment is made within three weeks of the assessment date or, if a different period is stated in the Contract Data, within the period stated. If a certified payment is late, or if a payment is late because the *Project Manager* has not issued a certificate which should be issued, interest is paid on the late payment. Interest is assessed from the date by which the late payment should have been made until the date when the late payment is made, and is included in the first assessment after the late payment is made.

51.3 If an amount due is corrected in a later certificate

- in relation to a mistake or a compensation event,
- because a payment was delayed by an unnecessary delay to a test or inspection done by the *Supervisor* or
- following a decision of the *Adjudicator* or the *tribunal*, or a recommendation of the Dispute Avoidance Board,

interest on the correcting amount is paid. Interest is assessed from the date when the incorrect amount was certified until the date when the changed amount is certified and is included in the assessment which includes the changed amount.

Of course, the final provision only applies if Option W3 is included. 

51.4 Interest is calculated on a daily basis at the *interest rate* and is compounded annually.

51.5 Any tax which the law requires a Party to pay to the other Party is added to any payment made under the contract.

Clause 51.5 is re-worded to keep the payment process such that one Party can pay the other Party, and not just the *Client* pays the *Contractor*. 

Defined Cost

52

52.1 All the *Contractor's* costs which are not included in the Defined Cost are treated as included in the Fee. Defined Cost includes only amounts calculated using rates and percentages stated in the Contract Data and other amounts at open market or competitively tendered prices with deductions for all discounts, rebates and taxes which can be recovered.

Final assessment

53

53.1 The *Project Manager* makes an assessment of the final amount due and certifies a final payment, if any is due, no later than

- four weeks after the *Supervisor* issues the Defects Certificate or
- thirteen weeks after the *Project Manager* issues a termination certificate.

The *Project Manager* gives the *Contractor* details of how the amount due has been assessed. The final payment is made within three weeks of the assessment or, if a different period is stated in the Contract Data, within the period stated.

53.2 If the *Project Manager* does not make this assessment within the time allowed, the *Contractor* may issue to the *Client* an assessment of the final amount due, giving details of how the final amount due has been assessed. If the *Client* agrees with this assessment, a final payment is made within two weeks of the assessment or, if a different period is stated in the Contract Data, within the period stated.

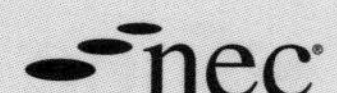

53.3 An assessment of the final amount due issued within the time stated in the contact is conclusive evidence of the final amount due under or in connection with the contract unless a Party takes the following actions.

If the contract includes Option W1, a Party

- refers a dispute about the assessment of the final amount due to the *Senior Representatives* within four weeks of the assessment being issued,
- refers any issues not agreed by the *Senior Representatives* to the *Adjudicator* within three weeks of the list of the issues not agreed being produced or when it should have been produced and
- refers to the *tribunal* its dissatisfaction with a decision of the *Adjudicator* as to the final assessment of the amount due within four weeks of the decision being made.

If the contract includes Option W2, a Party

- refers a dispute about the assessment of the final amount due to the *Senior Representatives* or to the *Adjudicator* within four weeks of the assessment being issued,
- refers any issues referred to but not agreed by the *Senior Representatives* to the *Adjudicator* within three weeks of the list of issues not agreed being produced or when it should have been produced and
- refers to the *tribunal* its dissatisfaction with a decision of the *Adjudicator* as to the final assessment of the amount due within four weeks of the decision being made.

If the contract includes Option W3, a Party

- refers a dispute about the assessment of the final amount due to the Dispute Avoidance Board and
- refers to the *tribunal* its dissatisfaction with the recommendation of the Dispute Avoidance Board within four weeks of the recommendation being made.

53.4 The assessment of the final amount due is changed to include

- any agreement the Parties reach and
- a decision of the *Adjudicator* or recommendation of the Dispute Avoidance Board which has not been referred to the *tribunal* within four weeks of that decision or recommendation.

A changed assessment becomes conclusive evidence of the final amount due under or in connection with the contract.

Clause 53 is a new clause that deals with closing down the financial aspects of the contract. Again, some might consider this to have been a surprising omission from previous versions, but whether it was a serious issue is debatable. The clause covers the dispute resolution procedure to be followed depending on which Option W clause is included in the contract and whether or not the *Project Manager* acts within the time allowed.

GLOSSARY OF NEW ECC4 TERMS	
Client	changed project role – formerly '*Employer*'
Dispute Avoidance Board	new entity, part of the revised dispute avoidance and resolution provisions (Option W3 only)
Senior Representatives	new entity, part of the revised dispute avoidance and resolution provisions (Options W1 and W2 only)
Scope	changed defined term – formerly 'Works Information'

6 Compensation events

Compensation events **60**

60.1 The following are compensation events.

(1) The *Project Manager* gives an instruction changing the Works Information except

- a change made in order to accept a Defect or
- a change to the Works Information provided by the *Contractor* for his design which is made either at his request or to comply with other Works Information provided by the *Employer*.

(2) The *Employer* does not allow access to and use of a part of the Site by the later of its *access date* and the date shown on the Accepted Programme.

(3) The *Employer* does not provide something which he is to provide by the date for providing it shown on the Accepted Programme.

(4) The *Project Manager* gives an instruction to stop or not to start any work or to change a Key Date.

(5) The *Employer* or Others

- do not work within the times shown on the Accepted Programme,
- do not work within the conditions stated in the Works Information or
- carry out work on the Site that is not stated in the Works Information.

(6) The *Project Manager* or the *Supervisor* does not reply to a communication from the *Contractor* within the period required by this contract.

(7) The *Project Manager* gives an instruction for dealing with an object of value or of historical or other interest found within the Site.

(8) The *Project Manager* or the *Supervisor* changes a decision which he has previously communicated to the *Contractor*.

(9) The *Project Manager* withholds an acceptance (other than acceptance of a quotation for acceleration or for not correcting a Defect) for a reason not stated in this contract.

(10) The *Supervisor* instructs the *Contractor* to search for a Defect and no Defect is found unless the search is needed only because the *Contractor* gave insufficient notice of doing work obstructing a required test or inspection.

(11) A test or inspection done by the *Supervisor* causes unnecessary delay.

(12) The *Contractor* encounters physical conditions which

- are within the Site,
- are not weather conditions and
- an experienced contractor would have judged at the Contract Date to have such a small chance of occurring that it would have been unreasonable for him to have allowed for them.

Only the difference between the physical conditions encountered and those for which it would have been reasonable to have allowed is taken into account in assessing a compensation event.

6. COMPENSATION EVENTS

Compensation events **60**

60.1 The following events are compensation events.

(1) The *Project Manager* gives an instruction changing the Scope except

- a change made in order to accept a Defect or
- a change to the Scope provided by the *Contractor* for its design which is made
  - at the *Contractor's* request or
  - in order to comply with the Scope provided by the *Client*.

(2) The *Client* does not allow access to and use of each part of the Site by the later of its *access date* and the date for access shown on the Accepted Programme.

(3) The *Client* does not provide something which it is to provide by the date shown in the Accepted Programme.

(4) The *Project Manager* gives an instruction to stop or not to start any work or to change a Key Date.

(5) The *Client* or Others

- do not work within the times shown on the Accepted Programme,
- do not work within the conditions stated in the Scope or
- carry out work on the Site that is not stated in the Scope.

(6) The *Project Manager* or the *Supervisor* does not reply to a communication from the *Contractor* within the period required by the contract.

(7) The *Project Manager* gives an instruction for dealing with an object of value or of historical or other interest found within the Site.

(8) The *Project Manager* or the *Supervisor* changes a decision which either has previously communicated to the *Contractor*.

(9) The *Project Manager* withholds an acceptance (other than acceptance of a quotation for acceleration or for not correcting a Defect) for a reason not stated in the contract.

(10) The *Supervisor* instructs the *Contractor* to search for a Defect and no Defect is found unless the search is needed only because the *Contractor* gave insufficient notice of doing work obstructing a required test or inspection.

(11) A test or inspection done by the *Supervisor* causes unnecessary delay.

(12) The *Contractor* encounters physical conditions which

- are within the Site,
- are not weather conditions and
- an experienced contractor would have judged at the Contract Date to have such a small chance of occurring that it would have been unreasonable to have allowed for them.

Only the difference between the physical conditions encountered and those for which it would have been reasonable to have allowed is taken into account in assessing a compensation event.

(13) A *weather measurement* is recorded

- within a calendar month,
- before the Completion Date for the whole of the *works* and
- at the place stated in the Contract Data

the value of which, by comparison with the *weather data*, is shown to occur on average less frequently than once in ten years.

Only the difference between the *weather measurement* and the weather which the *weather data* show to occur on average less frequently than once in ten years is taken into account in assessing a compensation event.

(14) An event which is an *Employer*'s risk stated in this contract.

(15) The *Project Manager* certifies take over of a part of the *works* before both Completion and the Completion Date.

(16) The *Employer* does not provide materials, facilities and samples for tests and inspections as stated in the Works Information.

(17) The *Project Manager* notifies a correction to an assumption which he has stated about a compensation event.

(18) A breach of contract by the *Employer* which is not one of the other compensation events in this contract.

(19) An event which

- stops the *Contractor* completing the *works* or
- stops the *Contractor* completing the *works* by the date shown on the Accepted Programme,

and which

- neither Party could prevent,
- an experienced contractor would have judged at the Contract Date to have such a small chance of occurring that it would have been unreasonable for him to have allowed for it and
- is not one of the other compensation events stated in this contract.

(13) A *weather measurement* is recorded

- within a calendar month,
- before the Completion Date for the whole of the *works* and
- the place stated in the Contract Data

the value of which, by comparison with the *weather data*, is shown to occur on average less frequently than once in ten years.

Only the difference between the *weather measurement* and the weather which the *weather data* show to occur on average less frequently than once in ten years is taken into account in assessing a compensation event.

(14) An event which is a *Client's* liability stated in these *conditions of contract*.

(15) The *Project Manager* certifies take over of a part of the *works* before both Completion and the Completion Date.

(16) The *Client* does not provide materials, facilities and samples for tests and inspections as stated in the Scope.

(17) The *Project Manager* notifies the *Contractor* of a correction to an assumption which the *Project Manager* stated about a compensation event.

(18) A breach of contract by the *Client* which is not one of the other compensation events in the contract.

(19) An event which

- stops the *Contractor* completing the whole of the *works* or
- stops the *Contractor* completing the whole of the *works* by the date for planned Completion shown on the Accepted Programme, and

which

- neither Party could prevent,
- an experienced contractor would have judged at the Contract Date to have such a small chance of occurring that it would have been unreasonable to have allowed for it and
- is not one of the other compensation events stated in the contract.

(20) The *Project Manager* notifies the *Contractor* that a quotation for a proposed instruction is not accepted.

(21) Additional compensation events stated in Contract Data part one.

Some minor drafting changes are made to the 19 compensation events previously listed, but with no intention to create any significantly different principles. New compensation event 60.1(20) deals (more fairly) with the position that the *Contractor* spends time and money on a proposed instruction that comes to nothing. New compensation event 60.1(21) reverts to the position in NEC2 Engineering and Construction Contract by pulling in additional compensation events as stated in the Contract Data part one (prepared by the *Client*). This sensible change should also help to reduce the number of Option Z clauses.

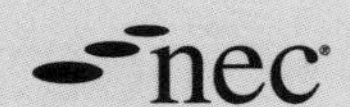

60.2 In judging the physical conditions for the purpose of assessing a compensation event, the *Contractor* is assumed to have taken into account

- the Site Information,
- publicly available information referred to in the Site Information,
- information obtainable from a visual inspection of the Site and
- other information which an experienced contractor could reasonably be expected to have or to obtain.

60.3 If there is an ambiguity or inconsistency within the Site Information (including the information referred to in it), the *Contractor* is assumed to have taken into account the physical conditions more favourable to doing the work.

Notifying compensation events **61**

61.1 For compensation events which arise from the *Project Manager* or the *Supervisor* giving an instruction, issuing a certificate, changing an earlier decision or correcting an assumption, the *Project Manager* notifies the *Contractor* of the compensation event at the time of that communication. He also instructs the *Contractor* to submit quotations, unless the event arises from a fault of the *Contractor* or quotations have already been submitted. The *Contractor* puts the instruction or changed decision into effect.

61.3 The *Contractor* notifies the *Project Manager* of an event which has happened or which he expects to happen as a compensation event if

- the *Contractor* believes that the event is a compensation event and
- the *Project Manager* has not notified the event to the *Contractor*.

If the *Contractor* does not notify a compensation event within eight weeks of becoming aware of the event, he is not entitled to a change in the Prices, the Completion Date or a Key Date unless the event arises from the *Project Manager* or the *Supervisor* giving an instruction, issuing a certificate, changing an earlier decision or correcting an assumption.

61.4 If the *Project Manager* decides that an event notified by the *Contractor*

- arises from a fault of the *Contractor*,
- has not happened and is not expected to happen,
- has no effect upon Defined Cost, Completion or meeting a Key Date or
- is not one of the compensation events stated in this contract

he notifies the *Contractor* of his decision that the Prices, the Completion Date and the Key Dates are not to be changed. If the *Project Manager* decides otherwise, he notifies the *Contractor* accordingly and instructs him to submit quotations.

The *Project Manager* notifies his decision to the *Contractor* and, if his decision is that the Prices, the Completion Date or the Key Dates are to be changed, instructs him to submit quotations before the end of either

- one week after the *Contractor*'s notification or
- a longer period to which the *Contractor* has agreed.

If the *Project Manager* does not notify his decision, the *Contractor* may notify the *Project Manager* of his failure. A failure by the *Project Manager* to reply within two weeks of this notification is treated as acceptance by the *Project Manager* that the event is a compensation event and an instruction to submit quotations.

60.2 In judging the physical conditions for the purpose of assessing a compensation event, the *Contractor* is assumed to have taken into account

- the Site Information,
- publicly available information referred to in the Site Information,
- information obtainable from a visual inspection of the Site and
- other information which an experienced contractor could reasonably be expected to have or to obtain.

60.3 If there is an ambiguity or inconsistency within the Site Information (including the information referred to in it), the *Contractor* is assumed to have taken into account the physical conditions more favourable to doing the work.

Notifying compensation events

61

61.1 For a compensation event which arises from the *Project Manager* or the *Supervisor* giving an instruction or notification, issuing a certificate or changing an earlier decision, the *Project Manager* notifies the *Contractor* of the compensation event at the time of that communication.

61.2 The *Project Manager* includes in the notification of a compensation event an instruction to the *Contractor* to submit quotations unless

- the event arises from a fault of the *Contractor* or
- the event has no effect upon Defined Cost, Completion or meeting a Key Date.

61.3 The *Contractor* notifies the *Project Manager* of an event which has happened or which is expected to happen as a compensation event if

- the *Contractor* believes that the event is a compensation event and
- the *Project Manager* has not notified the event to the *Contractor*.

If the *Contractor* does not notify a compensation event within eight weeks of becoming aware that the event has happened, the Prices, the Completion Date or a Key Date are not changed unless the event arises from the *Project Manager* or the *Supervisor* giving an instruction or notification, issuing a certificate or changing an earlier decision.

61.4 The *Project Manager* replies to the *Contractor's* notification of a compensation event within

- one week after the *Contractor's* notification or
- a longer period to which the *Contractor* has agreed.

If the event

- arises from a fault of the *Contractor*,
- has not happened and is not expected to happen,
- has not been notified within the timescales set out in these *conditions of contract*,
- has no effect upon Defined Cost, Completion or meeting a Key Date or
- is not one of the compensation events stated in the contract

the *Project Manager* notifies the *Contractor* that the Prices, the Completion Date and the Key Dates are not to be changed and states the reasons in the notification. Otherwise, the *Project Manager* notifies the *Contractor* that the event is a compensation event and includes in the notification an instruction to the *Contractor* to submit quotations.

If the *Project Manager* fails to reply to the *Contractor's* notification of a compensation event within the time allowed, the *Contractor* may notify the *Project Manager* of that failure. If the failure continues for a further two weeks after the *Contractor's* notification it is treated as acceptance by the *Project Manager* that the event is a compensation event and an instruction to submit quotations.

61.5 If the *Project Manager* decides that the *Contractor* did not give an early warning of the event which an experienced contractor could have given, he notifies this decision to the *Contractor* when he instructs him to submit quotations.

61.6 If the *Project Manager* decides that the effects of a compensation event are too uncertain to be forecast reasonably, he states assumptions about the event in his instruction to the *Contractor* to submit quotations. Assessment of the event is based on these assumptions. If any of them is later found to have been wrong, the *Project Manager* notifies a correction.

61.7 A compensation event is not notified after the *defects date*.

Quotations for compensation events **62**

62.1 After discussing with the *Contractor* different ways of dealing with the compensation event which are practicable, the *Project Manager* may instruct the *Contractor* to submit alternative quotations. The *Contractor* submits the required quotations to the *Project Manager* and may submit quotations for other methods of dealing with the compensation event which he considers practicable.

62.2 Quotations for compensation events comprise proposed changes to the Prices and any delay to the Completion Date and Key Dates assessed by the *Contractor*. The *Contractor* submits details of his assessment with each quotation. If the programme for remaining work is altered by the compensation event, the *Contractor* includes the alterations to the Accepted Programme in his quotation.

62.3 The *Contractor* submits quotations within three weeks of being instructed to do so by the *Project Manager*. The *Project Manager* replies within two weeks of the submission. His reply is

- an instruction to submit a revised quotation,
- an acceptance of a quotation,
- a notification that a proposed instruction will not be given or a proposed changed decision will not be made or
- a notification that he will be making his own assessment.

62.4 The *Project Manager* instructs the *Contractor* to submit a revised quotation only after explaining his reasons for doing so to the *Contractor*. The *Contractor* submits the revised quotation within three weeks of being instructed to do so.

62.5 The *Project Manager* extends the time allowed for

- the *Contractor* to submit quotations for a compensation event and
- the *Project Manager* to reply to a quotation

if the *Project Manager* and the *Contractor* agree to the extension before the submission or reply is due. The *Project Manager* notifies the extension that has been agreed to the *Contractor*.

62.6 If the *Project Manager* does not reply to a quotation within the time allowed, the *Contractor* may notify the *Project Manager* of his failure. If the *Contractor* submitted more than one quotation for the compensation event, he states in his notification which quotation he proposes is to be accepted. If the *Project Manager* does not reply to the notification within two weeks, and unless the quotation is for a proposed instruction or a proposed changed decision, the *Contractor*'s notification is treated as acceptance of the quotation by the *Project Manager*.

61.5 If the *Project Manager* decides that the *Contractor* did not give an early warning of the event which an experienced contractor could have given, the *Project Manager* states this in the instruction to the *Contractor* to submit quotations.

61.6 If the effects of a compensation event are too uncertain to be forecast reasonably, the *Project Manager* states assumptions about the compensation event in the instruction to the *Contractor* to submit quotations. Assessment of the event is based on these assumptions. If any of them is later found to have been wrong, the *Project Manager* notifies a correction.

61.7 A compensation event is not notified by the *Project Manager* or the *Contractor* after the issue of the Defects Certificate.

The drafting in clause 61 has had some modifications, but again the intent has been to make things clear, rather than to radically overhaul any existing NEC principles. 

Quotations for compensation events

62

62.1 After discussing with the *Contractor* different ways of dealing with the compensation event which are practicable, the *Project Manager* may instruct the *Contractor* to submit alternative quotations. The *Contractor* submits the required quotations to the *Project Manager* and may submit quotations for other methods of dealing with the compensation event which it considers practicable.

62.2 Quotations for a compensation event comprise proposed changes to the Prices and any delay to the Completion Date and Key Dates assessed by the *Contractor*. The *Contractor* submits details of the assessment with each quotation. If the programme for remaining work is altered by the compensation event, the *Contractor* includes the alterations to the Accepted Programme in the quotation.

62.3 The *Contractor* submits quotations within three weeks of being instructed to do so by the *Project Manager*. The *Project Manager* replies within two weeks of the submission. The reply is

- a notification of acceptance of the quotation,
- an instruction to submit a revised quotation or
- that the *Project Manager* will be making the assessment.

62.4 The *Project Manager* instructs the *Contractor* to submit a revised quotation only after explaining the reasons for doing so to the *Contractor*. The *Contractor* submits the revised quotation within three weeks of being instructed to do so.

62.5 The *Project Manager* extends the time allowed for

- the *Contractor* to submit quotations for a compensation event or
- the *Project Manager* to reply to a quotation

if the *Project Manager* and the *Contractor* agree to the extension before the submission or reply is due. The *Project Manager* informs the *Contractor* of the extension which has been agreed.

Note that the *Project Manager* now 'informs' rather than 'notifies' the *Contractor* of the extension. 

62.6 If the *Project Manager* does not reply to a quotation within the time allowed, the *Contractor* may notify the *Project Manager* of that failure. If the *Contractor* submitted more than one quotation for the compensation event, the notification states which quotation the *Contractor* proposes is to be used. If the failure continues for a further two weeks after the *Contractor's* notification it is treated as acceptance by the *Project Manager* of the quotation.

Clause 62 has had some modifications, but in all cases the changes have been to clarify the drafting. 

Assessing compensation events

63

63.1 The changes to the Prices are assessed as the effect of the compensation event upon

- the actual Defined Cost of the work already done,
- the forecast Defined Cost of the work not yet done and
- the resulting Fee.

If the compensation event arose from the *Project Manager* or the *Supervisor* giving an instruction, issuing a certificate, changing an earlier decision or correcting an assumption, the date which divides the work already done from the work not yet done is the date of that communication. In all other cases, the date is the date of the notification of the compensation event.

63.13 (Option B and D, 63.14 Option A, C, E and F) If the *Project Manager* and the *Contractor* agree, rates and lump sums may be used to assess a compensation event.

63.2 If the effect of a compensation event is to reduce the total Defined Cost, the Prices are not reduced except as stated in this contract.

63.3 A delay to the Completion Date is assessed as the length of time that, due to the compensation event, planned Completion is later than planned Completion as shown on the Accepted Programme. A delay to a Key Date is assessed as the length of time that, due to the compensation event, the planned date when the Condition stated for a Key Date will be met is later than the date shown on the Accepted Programme.

63.4 The rights of the *Employer* and the *Contractor* to changes to the Prices, the Completion Date and the Key Dates are their only rights in respect of a compensation event.

63.5 If the *Project Manager* has notified the *Contractor* of his decision that the *Contractor* did not give an early warning of a compensation event which an experienced contractor could have given, the event is assessed as if the *Contractor* had given early warning.

63.6 Assessment of the effect of a compensation event includes risk allowances for cost and time for matters which have a significant chance of occurring and are at the *Contractor*'s risk under this contract.

63.7 Assessments are based upon the assumptions that the *Contractor* reacts competently and promptly to the compensation event, that any Defined Cost and time due to the event are reasonably incurred and that the Accepted Programme can be changed.

Assessing compensation events

63

63.1 The change to the Prices is assessed as the effect of the compensation event upon

- the actual Defined Cost of the work done by the dividing date,
- the forecast Defined Cost of the work not done by the dividing date and
- the resulting Fee.

For a compensation event that arises from the *Project Manager* or the *Supervisor* giving an instruction or notification, issuing a certificate or changing an earlier decision, the dividing date is the date of that communication.

For other compensation events, the dividing date is the date of the notification of the compensation event.

With previous editions of the ECC, a question that has been much debated is at what point in the quotation should Defined Cost switch from actual to forecast? Is it when the quotation is carried out or should have been carried out, when the *Project Manager* gives an instruction or should have done, and so on. Often this has been referred to as the 'switch date'. The changes to clause 63.1 attempt to clarify this issue, by laying out a concept known as the 'dividing date', the date that divides actual from forecast Defined Cost.

63.2 The *Project Manager* and the *Contractor* may agree rates or lump sums to assess the change to the Prices.

New clause 63.2 is brought into the core clauses meaning that the equivalent clause in the main Options can be deleted, again saving on words.

63.3 If the effect of a compensation event is to reduce the total Defined Cost, the Prices are not reduced unless otherwise stated in these *conditions of contract*.

63.4 If the effect of a compensation event is to reduce the total Defined Cost and the event is

- a change to the Scope other than a change to the Scope provided by the *Client,* which the *Contractor* proposed and the *Project Manager* accepted or
- a correction to an assumption stated by the *Project Manager* for assessing an earlier compensation event

the Prices are reduced.

This clause originates from some of the main Options and has been drafted in a way that does not change the previous intentions.

63.5 A delay to the Completion Date is assessed as the length of time that, due to the compensation event, planned Completion is later than planned Completion as shown on the Accepted Programme current at the dividing date.

A delay to a Key Date is assessed as the length of time that, due to the compensation event, the planned date when the Condition stated for a Key Date will be met is later than the date shown on the Accepted Programme current at the dividing date.

When assessing delay only those operations which the *Contractor* has not completed and which are affected by the compensation event are changed.

Changes to this clause include a stipulation as to which version of the Accepted Programme is to be used to assess the delay in a compensation event assessment.

63.6 The rights of the *Client* and the *Contractor* to changes to the Prices, the Completion Date and the Key Dates are their only rights in respect of a compensation event.

63.7 If the *Project Manager* has stated in the instruction to submit quotations that the *Contractor* did not give an early warning of the event which an experienced contractor could have given, the compensation event is assessed as if the *Contractor* had given the early warning.

63.8 A compensation event which is an instruction to change the Works Information in order to resolve an ambiguity or inconsistency is assessed as if the Prices, the Completion Date and the Key Dates were for the interpretation most favourable to the Party which did not provide the Works Information.

63.9 If a change to the Works Information makes the description of the Condition for a Key Date incorrect, the *Project Manager* corrects the description. This correction is taken into account in assessing the compensation event for the change to the Works Information.

The *Project Manager*'s assessments **64**

64.1 The *Project Manager* assesses a compensation event

- if the *Contractor* has not submitted a quotation and details of his assessment within the time allowed,
- if the *Project Manager* decides that the *Contractor* has not assessed the compensation event correctly in a quotation and he does not instruct the *Contractor* to submit a revised quotation,
- if, when the *Contractor* submits quotations for a compensation event, he has not submitted a programme or alterations to a programme which this contract requires him to submit or
- if, when the *Contractor* submits quotations for a compensation event, the *Project Manager* has not accepted the *Contractor*'s latest programme for one of the reasons stated in this contract.

64.2 The *Project Manager* assesses a compensation event using his own assessment of the programme for the remaining work if

- there is no Accepted Programme or
- the *Contractor* has not submitted a programme or alterations to a programme for acceptance as required by this contract.

64.3 The *Project Manager* notifies the *Contractor* of his assessment of a compensation event and gives him details of it within the period allowed for the *Contractor*'s submission of his quotation for the same event. This period starts when the need for the *Project Manager*'s assessment becomes apparent.

64.4 If the *Project Manager* does not assess a compensation event within the time allowed, the *Contractor* may notify the *Project Manager* of his failure. If the *Contractor* submitted more than one quotation for the compensation event, he states in his notification which quotation he proposes is to be accepted. If the *Project Manager* does not reply within two weeks of this notification the notification is treated as acceptance of the *Contractor's* quotation by the *Project Manager*.

63.8 The assessment of the effect of a compensation event includes risk allowances for cost and time for matters which have a significant chance of occurring and are not compensation events.

63.9 The assessment of the effect of a compensation event is based upon the assumptions that the *Contractor* reacts competently and promptly to the event and that any Defined Cost and time due to the event are reasonably incurred.

63.10 A compensation event which is an instruction to change the Scope in order to resolve an ambiguity or inconsistency is assessed as if the Prices, the Completion Date and the Key Dates were for the interpretation most favourable to the Party which did not provide the Scope.

63.11 If a change to the Scope makes the description of the Condition for a Key Date incorrect, the *Project Manager* corrects the description. This correction is taken into account in assessing the compensation event for the change to the Scope.

The *Project Manager's* assessments

64

64.1 The *Project Manager* assesses a compensation event

- if the *Contractor* has not submitted the quotation and details of its assessment within the time allowed,
- if the *Project Manager* decides that the *Contractor* has not assessed the compensation event correctly in the quotation and has not instructed the *Contractor* to submit a revised quotation,
- if, when the *Contractor* submits quotations for the compensation event, it has not submitted a programme or alterations to a programme which the contract requires it to submit or
- if, when the *Contractor* submits quotations for the compensation event, the *Project Manager* has not accepted the *Contractor's* latest programme for one of the reasons stated in the contract.

64.2 The *Project Manager* assesses the programme for the remaining work and uses it in the assessment of a compensation event if

- there is no Accepted Programme,
- the *Contractor* has not submitted a programme or alterations to a programme for acceptance as required by the contract or
- the *Project Manager* has not accepted the *Contractor's* latest programme for one of the reasons stated in the contract.

64.3 The *Project Manager* notifies the *Contractor* of the assessment of a compensation event and gives details of the assessment within the period allowed for the *Contractor's* submission of its quotation for the same compensation event. This period starts when the need for the *Project Manager's* assessment becomes apparent.

64.4 If the *Project Manager* does not assess a compensation event within the time allowed, the *Contractor* may notify the *Project Manager* of that failure. If the *Contractor* submitted more than one quotation for the compensation event, the notification states which quotation the *Contractor* proposes is to be used. If the failure continues for a further two weeks after the *Contractor's* notification it is treated as acceptance by the *Project Manager* of the quotation.

Clause 64 has some minor modifications, but without any intention to change any of the principles.

This new clause brings together all the provisions for a proposed instruction, from start to finish, rather than having them scattered throughout the compensation event process. The result is a much tidier drafting approach. There is also a new compensation event (clause 60.1(20)) that applies if a proposed instruction does not go ahead.

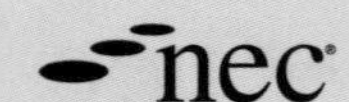

61.2 The *Project Manager* may instruct the *Contractor* to submit quotations for a proposed instruction or a proposed changed decision. The *Contractor* does not put a proposed instruction or a proposed changed decision into effect.

Implementing compensation events **65**

65.1 A compensation event is implemented when

- the *Project Manager* notifies his acceptance of the *Contractor*'s quotation,
- the *Project Manager* notifies the *Contractor* of his own assessment or
- a *Contractor*'s quotation is treated as having been accepted by the *Project Manager*.

65.2 The assessment of a compensation event is not revised if a forecast upon which it is based is shown by later recorded information to have been wrong.

Proposed instructions

65

65.1 The *Project Manager* may instruct the *Contractor* to submit a quotation for a proposed instruction. The *Project Manager* states in the instruction the date by which the proposed instruction may be given. The *Contractor* does not put a proposed instruction into effect.

65.2 The *Contractor* submits quotations for a proposed instruction within three weeks of being instructed to do so by the *Project Manager*. The quotation is assessed as a compensation event. The *Project Manager* replies to the *Contractor's* quotation by the date when the proposed instruction may be given. The reply is

- an instruction to submit a revised quotation including the reasons for doing so,
- the issue of the instruction together with a notification of the instruction as a compensation event and acceptance of the quotation or
- a notification that the quotation is not accepted.

If the *Project Manager* does not reply to the quotation within the time allowed, the quotation is not accepted.

65.3 If the quotation is not accepted, the *Project Manager* may issue the instruction, notify the instruction as a compensation event and instruct the *Contractor* to submit a quotation.

This clause is renumbered to allow for clause 65. There are again no real changes in principle here, just an attempt to make the wording even more clear. In clause 66.3, such a revision to the Prices could take place, for example, through the use of Option W1, W2 or W3. The redrafting in clause 66.2 has allowed this provision to be included as a core clause, and previous clause 65.3 of Options E and F and clause 65.4 of Options A to D to be removed, simplifying the drafting.

Implementing compensation events

66

66.1 A compensation event is implemented when

- the *Project Manager* notifies acceptance of the *Contractor's* quotation,
- the *Project Manager* notifies the *Contractor* of an assessment made by the *Project Manager* or
- a *Contractor's* quotation is treated as having been accepted by the *Project Manager*.

66.2 When a compensation event is implemented the Prices, the Completion Date and the Key Dates are changed accordingly.

66.3 The assessment of an implemented compensation event is not revised except as stated in these *conditions of contract*.

GLOSSARY OF NEW ECC4 TERMS	
Client	changed project role – formerly *'Employer'*
dividing date	new contract term – used in the assessment of compensation events
Scope	changed defined term – formerly 'Works Information'

7 Title

The *Employer's* title to Plant and Materials

70

70.1 Whatever title the *Contractor* has to Plant and Materials which is outside the Working Areas passes to the *Employer* if the *Supervisor* has marked it as for this contract.

70.2 Whatever title the *Contractor* has to Plant and Materials passes to the *Employer* if it has been brought within the Working Areas. The title to Plant and Materials passes back to the *Contractor* if it is removed from the Working Areas with the *Project Manager*'s permission.

Marking Equipment, Plant and Materials outside the Working Areas

71

71.1 The *Supervisor* marks Equipment, Plant and Materials which are outside the Working Areas if

- this contract identifies them for payment and
- the *Contractor* has prepared them for marking as the Works Information requires.

Removing Equipment

72

72.1 The *Contractor* removes Equipment from the Site when it is no longer needed unless the *Project Manager* allows it to be left in the *works*.

Objects and materials within the Site

73

73.1 The *Contractor* has no title to an object of value or of historical or other interest within the Site. The *Contractor* notifies the *Project Manager* when such an object is found and the *Project Manager* instructs the *Contractor* how to deal with it. The *Contractor* does not move the object without instructions.

73.2 The *Contractor* has title to materials from excavation and demolition only as stated in the Works Information.

7. TITLE

The *Client's* title to Plant and Materials 70

70.1 Whatever title the *Contractor* has to Plant and Materials which are outside the Working Areas passes to the *Client* if the *Supervisor* has marked them as for the contract.

70.2 Whatever title the *Contractor* has to Plant and Materials passes to the *Client* if they have been brought within the Working Areas. The title to Plant and Materials passes back to the *Contractor* if they are removed from the Working Areas with the *Project Manager's* permission.

Marking Equipment, Plant and Materials outside the Working Areas 71

71.1 The *Supervisor* marks Equipment, Plant and Materials which are outside the Working Areas if

- the contract identifies them for payment and
- the *Contractor* has prepared them for marking as the Scope requires.

Removing Equipment 72

72.1 The *Contractor* removes Equipment from the Site when it is no longer needed unless the *Project Manager* allows it to be left in the *works*.

Objects and materials within the Site 73

73.1 The *Contractor* has no title to an object of value or of historical or other interest within the Site. The *Contractor* informs the *Project Manager* when such an object is found and the *Project Manager* instructs the *Contractor* how to deal with it. The *Contractor* does not move the object without instructions.

Note that the *Contractor* now 'informs' rather than 'notifies' the *Project Manager* on finding an object.

73.2 The *Contractor* has title to materials from excavation and demolition unless the Scope states otherwise.

Clause 73.2 now reverses the previous approach on title to materials from excavation and demolition. The default is now that title of such goes to the *Contractor*, which is arguably a much more sensible approach. The issue will still need to be addressed carefully when preparing tender documents.

The *Contractor's* use of material 74

74.1 The *Contractor* has the right to use material provided by the *Client* only to Provide the Works. The *Contractor* may make this right available to a Subcontractor.

Clause 74.1 is a new provision dealing with the right to use material provided by the *Client*. It cannot be used other than to Provide the Works and the *Contractor* has the right to extend this to any Subcontractor.

GLOSSARY OF NEW ECC4 TERMS	
Client	changed project role – formerly '*Employer*'
Scope	changed defined term – formerly 'Works Information'

8 Risks and insurance

***Employer*'s risks**

80

80.1 The following are *Employer*'s risks.

- Claims, proceedings, compensation and costs payable which are due to
  - use or occupation of the Site by the *works* or for the purpose of the *works* which is the unavoidable result of the *works*,
  - negligence, breach of statutory duty or interference with any legal right by the *Employer* or by any person employed by or contracted to him except the *Contractor* or
  - a fault of the *Employer* or a fault in his design.
- Loss of or damage to Plant and Materials supplied to the *Contractor* by the *Employer*, or by Others on the *Employer*'s behalf, until the *Contractor* has received and accepted them.
- Loss of or damage to the *works*, Plant and Materials due to
  - war, civil war, rebellion, revolution, insurrection, military or usurped power,
  - strikes, riots and civil commotion not confined to the *Contractor*'s employees or
  - radioactive contamination.
- Loss of or wear or damage to the parts of the *works* taken over by the *Employer*, except loss, wear or damage occurring before the issue of the Defects Certificate which is due to
  - a Defect which existed at take over,
  - an event occurring before take over which was not itself an *Employer*'s risk or
  - the activities of the *Contractor* on the Site after take over.
- Loss of or wear or damage to the *works* and any Equipment, Plant and Materials retained on the Site by the *Employer* after a termination, except loss, wear or damage due to the activities of the *Contractor* on the Site after the termination.
- Additional *Employer*'s risks stated in the Contract Data.

The *Contractor*'s risks

81

81.1 From the *starting date* until the Defects Certificate has been issued, the risks which are not carried by the *Employer* are carried by the *Contractor*.

Repairs

82

82.1 Until the Defects Certificate has been issued and unless otherwise instructed by the *Project Manager*, the *Contractor* promptly replaces loss of and repairs damage to the *works*, Plant and Materials.

8. LIABILITIES AND INSURANCE

***Client's* liabilities** **80**

80.1 The following are *Client's* liabilities.

- Claims and proceedings from Others and compensation and costs payable to Others which are due to
  - use or occupation of the Site by the *works* or for the purpose of the *works* which is the unavoidable result of the *works*,
  - negligence, breach of statutory duty or interference with any legal right by the *Client* or by any person employed by or contracted to it except the *Contractor*.
- A fault of the *Client* or any person employed by or contracted to it, except the *Contractor*.
- A fault in the design contained in
  - the Scope provided by the *Client* or
  - an instruction from the *Project Manager* changing the Scope.
- Loss of or damage to Plant and Materials supplied to the *Contractor* by the *Client*, or by Others on the *Client's* behalf, until the *Contractor* has received and accepted them.
- Loss of or damage to the *works*, Plant and Materials due to
  - war, civil war, rebellion, revolution, insurrection, military or usurped power,
  - strikes, riots and civil commotion not confined to the *Contractor's* employees or
  - radioactive contamination.
- Loss of or damage to the parts of the *works* taken over by the *Client*, except loss or damage occurring before the issue of the Defects Certificate which is due to
  - a Defect which existed at take over,
  - an event occurring before take over which was not itself a *Client's* liability or
  - the activities of the *Contractor* on the Site after take over.
- Loss of or damage to the *works* and any Equipment, Plant and Materials retained on the Site by the *Client* after a termination, except loss or damage due to the activities of the *Contractor* on the Site after the termination.
- Loss of or damage to property owned or occupied by the *Client* other than the *works*, unless the loss or damage arises from or in connection with the *Contractor* Providing the Works.
- Additional *Client's* liabilities stated in the Contract Data.

The main change in this section is the use of the word 'liability' rather than 'risk', the former having being the intention of the drafters in previous editions anyway. This change also helps users by reducing the amount of times the word 'risk' is used, when another would probably be better.

Indemnity

83

83.1 Each Party indemnifies the other against claims, proceedings, compensation and costs due to an event which is at his risk.

83.2 The liability of each Party to indemnify the other is reduced if events at the other Party's risk contributed to the claims, proceedings, compensation and costs. The reduction is in proportion to the extent that events which were at the other Party's risk contributed, taking into account each Party's responsibilities under this contract.

Insurance cover

84

84.1 The *Contractor* provides the insurances stated in the Insurance Table except any insurance which the *Employer* is to provide as stated in the Contract Data. The *Contractor* provides additional insurances as stated in the Contract Data.

84.2 The insurances are in the joint names of the Parties and provide cover for events which are at the *Contractor*'s risk from the *starting date* until the Defects Certificate or a termination certificate has been issued.

Contractor's liabilities

81

81.1 The following are *Contractor's* liabilities unless they are stated as being *Client's* liabilities.

- Claims and proceedings from Others and compensation and costs payable to Others which arise from or in connection with the *Contractor* Providing the Works.
- Loss of or damage to the *works*, Plant and Materials and Equipment.
- Loss of or damage to property owned or occupied by the *Client* other than the *works*, which arises from or in connection with the *Contractor* Providing the Works.
- Death or bodily injury to the employees of the *Contractor*.

Clause 81.1 now takes a different drafting approach to that used in ECC3. Rather than the 'all others are yours' approach used previously, specific *Contractor's* liabilities are now stated. This will perhaps help contractors to get competitive insurance.

Recovery of costs

82

82.1 Any cost which the *Client* has paid or will pay as a result of an event for which the *Contractor* is liable is paid by the *Contractor*.

82.2 Any cost which the *Contractor* has paid or will pay to Others as a result of an event for which the *Client* is liable is paid by the *Client*.

82.3 The right of a Party to recover these costs is reduced if an event for which they were liable contributed to the costs. The reduction is in proportion to the extent that the event for which that Party is liable contributed, taking into account each Party's responsibilities under the contract.

By changing the title from 'Indemnity' – which is arguably not a term everyone understands – to 'Recovery of costs', the drafters have made clear exactly what the clause is all about. Note that the previous clause 82 (repairs) has been deleted.

Insurance cover

83

83.1 The *Client* provides the insurances which the *Client* is to provide as stated in the Contract Data.

83.2 The *Contractor* provides the insurances stated in the Insurance Table except any insurance which the *Client* is to provide as stated in the Contract Data. The *Contractor* provides additional insurances as stated in the Contract Data.

83.3 The insurances in the Insurance Table are in the joint names of the Parties except the fourth insurance stated. The insurances provide cover for events which are the *Contractor's* liability from the *starting date* until the Defects Certificate or a termination certificate has been issued.

There was some kickback over joint names insurance cover, which has been recognised in the changes made to the fourth insurance in the Insurance Table.

INSURANCE TABLE

Insurance against	Minimum amount of cover or minimum limit of indemnity
Loss of or damage to the *works*, Plant and Materials	The replacement cost, including the amount stated in the Contract Data for the replacement of any Plant and Materials provided by the *Employer*
Loss of or damage to Equipment	The replacement cost
Liability for loss of or damage to property (except the *works*, Plant and Materials and Equipment) and liability for bodily injury to or death of a person (not an employee of the *Contractor*) caused by activity in connection with this contract	The amount stated in the Contract Data for any one event with cross liability so that the insurance applies to the Parties separately
Liability for death of or bodily injury to employees of the *Contractor* arising out of and in the course of their employment in connection with this contract	The greater of the amount required by the applicable law and the amount stated in the Contract Data for any one event

Insurance policies **85**

85.1 Before the *starting date* and on each renewal of the insurance policy until the *defects date*, the *Contractor* submits to the *Project Manager* for acceptance certificates which state that the insurance required by this contract is in force. The certificates are signed by the *Contractor*'s insurer or insurance broker. A reason for not accepting the certificates is that they do not comply with this contract.

85.2 Insurance policies include a waiver by the insurers of their subrogation rights against directors and other employees of every insured except where there is fraud.

85.3 The Parties comply with the terms and conditions of the insurance policies.

85.4 Any amount not recovered from an insurer is borne by the *Employer* for events which are at his risk and by the *Contractor* for events which are at his risk.

If the *Contractor* does not insure **86**

86.1 The *Employer* may insure a risk which this contract requires the *Contractor* to insure if the *Contractor* does not submit a required certificate. The cost of this insurance to the *Employer* is paid by the *Contractor*.

Insurance by the *Employer* **87**

87.1 The *Project Manager* submits policies and certificates for insurances provided by the *Employer* to the *Contractor* for acceptance before the *starting date* and afterwards as the *Contractor* instructs. The *Contractor* accepts the policies and certificates if they comply with this contract.

87.2 The *Contractor*'s acceptance of an insurance policy or certificate provided by the *Employer* does not change the responsibility of the *Employer* to provide the insurances stated in the Contract Data.

87.3 The *Contractor* may insure a risk which this contract requires the *Employer* to insure if the *Employer* does not submit a required policy or certificate. The cost of this insurance to the *Contractor* is paid by the *Employer*.

INSURANCE TABLE	
INSURANCE AGAINST	**MINIMUM AMOUNT OF COVER**
Loss of or damage to the *works*, Plant and Materials	The replacement cost, including the amount stated in the Contract Data for the replacement of any Plant and Materials provided by the *Client*
Loss of or damage to Equipment	The replacement cost
Loss of or damage to property (except the *works*, Plant and Materials and Equipment) and liability for bodily injury to or death of a person (not an employee of the *Contractor*) arising from or in connection with the *Contractor* Providing the Works.	The amount stated in the Contract Data for any one event with cross liability so that the insurance applies to the Parties separately
Death of or bodily injury to employees of the *Contractor* arising out of and in the course of their employment in connection with the contract	The greater of the amount required by the applicable law and the amount stated in the Contract Data for any one event

Insurance policies 84

84.1 Before the *starting date* and on each renewal of the insurance policy until the *defects date*, the *Contractor* submits to the *Project Manager* for acceptance certificates which state that the insurance required by the contract is in force. The certificates are signed by the *Contractor's* insurer or insurance broker. The *Project Manager* accepts the certificates if the insurance complies with the contract and if the insurer's commercial position is strong enough to carry the insured liabilities.

84.2 Insurance policies include a waiver by the insurers of their subrogation rights against the Parties and the directors and other employees of every insured except where there is fraud.

84.3 The Parties comply with the terms and conditions of the insurance policies to which they are a party.

If the *Contractor* does not insure 85

85.1 The *Client* may insure an event or liability which the contract requires the *Contractor* to insure if the *Contractor* does not submit a required certificate. The cost of this insurance to the *Client* is paid by the *Contractor*.

Insurance by the *Client* 86

86.1 The *Project Manager* submits certificates for insurance provided by the *Client* to the *Contractor* for acceptance before the *starting date* and afterwards as the *Contractor* instructs. The *Contractor* accepts the certificates if the insurance complies with the contract and if the insurer's commercial position is strong enough to carry the insured liabilities.

86.2 The *Contractor's* acceptance of an insurance certificate provided by the *Client* does not change the responsibility of the *Client* to provide the insurances stated in the Contract Data.

86.3 The *Contractor* may insure an event or liability which the contract requires the *Client* to insure if the *Client* does not submit a required certificate. The cost of this insurance to the *Contractor* is paid by the *Client*.

GLOSSARY OF NEW ECC4 TERMS	
Client	changed project role – formerly '*Employer*'
Scope	changed defined term – formerly 'Works Information'
liabilities	changed contract term – formerly 'risks'
Liabilities and insurance	changed section title – formerly 'Risks and insurance'
Recovery of costs	changed term – formerly 'indemnity'

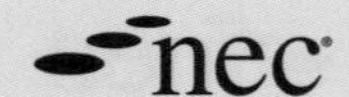

CORE CLAUSES

MAIN OPTION CLAUSES

SECONDARY OPTION CLAUSES

COST COMPONENTS

9 Termination

Termination

90

90.1 If either Party wishes to terminate the *Contractor*'s obligation to Provide the Works he notifies the *Project Manager* and the other Party giving details of his reason for terminating. The *Project Manager* issues a termination certificate to both Parties promptly if the reason complies with this contract.

90.2 The *Contractor* may terminate only for a reason identified in the Termination Table. The *Employer* may terminate for any reason. The procedures followed and the amounts due on termination are in accordance with the Termination Table.

TERMINATION TABLE

Terminating Party	Reason	Procedure	Amount due
The *Employer*	A reason other than R1–R21	P1 and P2	A1, A2 and A4
	R1–R15 or R18	P1, P2 and P3	A1 and A3
	R17 or R20	P1 and P3	A1 and A2
	R21	P1 and P4	A1 and A2
The *Contractor*	R1–R10, R16 or R19	P1 and P4	A1, A2 and A4
	R17 or R20	P1 and P4	A1 and A2

90.3 The procedures for termination are implemented immediately after the *Project Manager* has issued a termination certificate.

90.4 Within thirteen weeks of termination, the *Project Manager* certifies a final payment to or from the *Contractor* which is the *Project Manager*'s assessment of the amount due on termination less the total of previous payments. Payment is made within three weeks of the *Project Manager*'s certificate.

90.5 After a termination certificate has been issued, the *Contractor* does no further work necessary to Provide the Works.

9. TERMINATION

Termination **90**

90.1 If either Party wishes to terminate the *Contractor's* obligation to Provide the Works they notify the *Project Manager* and the other Party giving details of the reason for terminating. The *Project Manager* issues a termination certificate promptly if the reason complies with the contract.

90.2 A Party may terminate for a reason identified in the Termination Table. The procedures followed and the amounts due on termination are in accordance with the Termination Table.

TERMINATION TABLE			
TERMINATING PARTY	REASON	PROCEDURE	AMOUNT DUE
The *Client*	R1–R15, R18 or R22	P1, P2 and P3	A1 and A3
	R17 or R20	P1 and P3	A1 and A2
	R21	P1 and P4	A1 and A2
The *Contractor*	R1–R10, R16 or R19	P1 and P4	A1, A2 and A4
	R17 or R20	P1 and P4	A1 and A2

90.3 The procedures for termination are implemented immediately after the *Project Manager* has issued a termination certificate.

If the *Client* terminates for one of reasons R1 to R15, R18 or R22 and a certified payment has not been made at the date of the termination certificate, the *Client* need not make the certified payment unless these *conditions of contract* state otherwise.

90.4 After a termination certificate has been issued, the *Contractor* does no further work necessary to Provide the Works.

Reasons for termination **91**

91.1 Either Party may terminate if the other Party has done one of the following or its equivalent.

- If the other Party is an individual and has
  - presented his petition for bankruptcy (R1),
  - had a bankruptcy order made against him (R2),
  - had a receiver appointed over his assets (R3) or
  - made an arrangement with his creditors (R4).
- If the other Party is a company or partnership and has
  - had a winding-up order made against it (R5),
  - had a provisional liquidator appointed to it (R6),
  - passed a resolution for winding-up (other than in order to amalgamate or reconstruct) (R7),
  - had an administration order made against it (R8),
  - had a receiver, receiver and manager, or administrative receiver appointed over the whole or a substantial part of its undertaking or assets (R9) or
  - made an arrangement with its creditors (R10).

91.2 The *Employer* may terminate if the *Project Manager* has notified that the *Contractor* has defaulted in one of the following ways and not put the default right within four weeks of the notification.

- Substantially failed to comply with his obligations (R11).
- Not provided a bond or guarantee which this contract requires (R12).
- Appointed a Subcontractor for substantial work before the *Project Manager* has accepted the Subcontractor (R13).

91.3 The *Employer* may terminate if the *Project Manager* has notified that the *Contractor* has defaulted in one of the following ways and not stopped defaulting within four weeks of the notification.

- Substantially hindered the *Employer* or Others (R14).
- Substantially broken a health or safety regulation (R15).

91.4 The *Contractor* may terminate if the *Employer* has not paid an amount due under the contract within eleven weeks of the date that it should have been paid (R16).

91.5 Either Party may terminate if the Parties have been released under the law from further performance of the whole of this contract (R17).

Reasons for termination

91

91.1 Either Party may terminate if the other Party has done one of the following or its equivalent.

- If the other Party is an individual and has
  - presented an application for bankruptcy (R1),
  - had a bankruptcy order made against it (R2),
  - had a receiver appointed over its assets (R3) or
  - made an arrangement with its creditors (R4).
- If the other Party is a company or partnership and has
  - had a winding-up order made against it (R5),
  - had a provisional liquidator appointed to it (R6),
  - passed a resolution for winding-up (other than in order to amalgamate or reconstruct) (R7),
  - had an administration order made against it or had an administrator appointed over it (R8),
  - had a receiver, receiver and manager, or administrative receiver appointed over the whole or a substantial part of its undertaking or assets (R9) or
  - made an arrangement with its creditors (R10).

91.2 The *Client* may terminate if the *Project Manager* has notified that the *Contractor* has not put one of the following defaults right within four weeks of the date when the *Project Manager* notified the *Contractor* of the default.

- Substantially failed to comply with its obligations (R11).
- Not provided a bond or guarantee which the contract requires (R12).
- Appointed a Subcontractor for substantial work before the *Project Manager* has accepted the Subcontractor (R13).

91.3 The *Client* may terminate if the *Project Manager* has notified that the *Contractor* has not stopped one of the following defaults within four weeks of the date when the *Project Manager* notified the *Contractor* of the default.

- Substantially hindered the *Client* or Others (R14).
- Substantially broken a health or safety regulation (R15).

91.4 The *Contractor* may terminate if the *Client* has not paid an amount due under the contract within thirteen weeks of the date that the *Contractor* should have been paid (R16).

The period allowed for the *Client's* non-payment to continue is increased from 11 weeks to 13 weeks. 

91.5 Either Party may terminate if the Parties have been released under the law from further performance of the whole of the contract (R17).

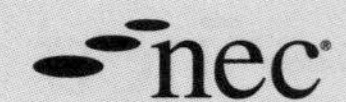

91.6 If the *Project Manager* has instructed the *Contractor* to stop or not to start any substantial work or all work and an instruction allowing the work to re-start or start has not been given within thirteen weeks,

- the Employer may terminate if the instruction was due to a default by the *Contractor* (R18),
- the *Contractor* may terminate if the instruction was due to a default by the Employer (R19) and
- either Party may terminate if the instruction was due to any other reason (R20).

91.7 The Employer may terminate if an event occurs which

- stops the *Contractor* completing the *works* or
- stops the *Contractor* completing the *works* by the date shown on the Accepted Programme and is forecast to delay Completion by more than 13 weeks,

and which

- neither Party could prevent and
- an experienced contractor would have judged at the Contract Date to have such a small chance of occurring that it would have been unreasonable for him to have allowed for it (R21).

Procedures on termination

92

92.1 On termination, the Employer may complete the *works* and may use any Plant and Materials to which he has title (P1).

92.2 The procedure on termination also includes one or more of the following as set out in the Termination Table.

P2 The Employer may instruct the *Contractor* to leave the Site, remove any Equipment, Plant and Materials from the Site and assign the benefit of any subcontract or other contract related to performance of this contract to the Employer.

P3 The Employer may use any Equipment to which the *Contractor* has title to complete the *works*. The *Contractor* promptly removes the Equipment from Site when the *Project Manager* notifies him that the Employer no longer requires it to complete the *works*.

P4 The *Contractor* leaves the Working Areas and removes the Equipment.

91.6 If the *Project Manager* has instructed the *Contractor* to stop or not to start any substantial work or all work and an instruction allowing the work to re-start or start or removing work from the Scope has not been given within thirteen weeks,

- the *Client* may terminate if the instruction was due to a default by the *Contractor* (R18),
- the *Contractor* may terminate if the instruction was due to a default by the *Client* (R19) and
- either Party may terminate if the instruction was due to any other reason (R20).

91.7 The *Client* may terminate if an event occurs which

- stops the *Contractor* completing the whole of the *works* or
- stops the *Contractor* completing the whole of the *works* by the date for planned Completion shown on the Accepted Programme and is forecast to delay Completion of the whole of the *works* by more than thirteen weeks,

and which

- neither Party could prevent and
- an experienced contractor would have judged at the Contract Date to have such a small chance of occurring that it would have been unreasonable to have allowed for it (R21).

The slight changes in the drafting of the prevention clause clarify that the problem must affect the whole of the *works*. 

91.8 The *Client* may terminate if the *Contractor* does a Corrupt Act, unless it was done by a Subcontractor or supplier and the *Contractor*

- was not and should not have been aware of the Corrupt Act or
- informed the *Project Manager* of the Corrupt Act and took action to stop it as soon as the *Contractor* became aware of it (R22).

This new reason is added to give the *Client* the right to terminate in the event the *Contractor* performs a Corrupt Act. It reflects the position many clients wish to take on this important matter. 

Procedures on termination

92

92.1 On termination, the *Client* may complete the *works* and may use any Plant and Materials to which it has title (P1).

92.2 The procedure on termination also includes one or more of the following as set out in the Termination Table.

P2 The *Client* may instruct the *Contractor* to leave the Site, remove any Equipment, Plant and Materials from the Site and assign the benefit of any subcontract or other contract related to performance of the contract to the *Client*.

P3 The *Client* may use any Equipment to which the *Contractor* has title to complete the *works*. The *Contractor* promptly removes the Equipment from Site when the *Project Manager* informs the *Contractor* that the *Client* no longer requires it to complete the *works*.

Note that the *Project Manager* now 'informs' rather than 'notifies' the *Contractor* of the *Client's* decision. 

P4 The *Contractor* leaves the Working Areas and removes the Equipment.

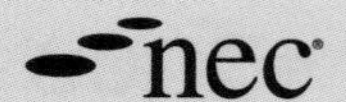

Payment on termination **93**

93.1 The amount due on termination includes (A1)

- an amount due assessed as for normal payments,
- the Defined Cost for Plant and Materials
  - within the Working Areas or
  - to which the *Employer* has title and of which the *Contractor* has to accept delivery,
- other Defined Cost reasonably incurred in expectation of completing the whole of the *works*,
- any amounts retained by the *Employer* and
- a deduction of any un-repaid balance of an advanced payment.

93.2 The amount due on termination also includes one or more of the following as set out in the Termination Table.

A2 The forecast Defined Cost of removing the Equipment.

A3 A deduction of the forecast of the additional cost to the *Employer* of completing the whole of the *works*.

A4 The *direct fee percentage* applied to

- for Options A, B, C and D, any excess of the total of the Prices at the Contract Date over the Price for Work Done to Date or
- for Options E and F, any excess of the first forecast of the Defined Cost for the *works* over the Price for Work Done to Date less the Fee.

Payment on termination

93

93.1 The amount due on termination includes (A1)

- an amount due assessed as for normal payments,
- the Defined Cost for Plant and Materials
  - within the Working Areas or
  - to which the *Client* has title and of which the *Contractor* has to accept delivery,
- other Defined Cost reasonably incurred in expectation of completing the whole of the *works*,
- any amounts retained by the *Client* and
- a deduction of any un-repaid balance of an advanced payment.

93.2 The amount due on termination also includes one or more of the following as set out in the Termination Table.

A2 The forecast Defined Cost of removing the Equipment.

A3 A deduction of the forecast of the additional cost to the *Client* of completing the whole of the *works*.

A4 The *fee percentage* applied to

- for Options A, B, C and D, any excess of the total of the Prices at the Contract Date over the Price for Work Done to Date or
- for Options E and F, any excess of the first forecast of the Defined Cost for the *works* over the Price for Work Done to Date less the Fee.

GLOSSARY OF NEW ECC4 TERMS	
Client	changed project role – formerly '*Employer*'
Corrupt Act	new defined term – defined in clause 11.2(5)

CORE CLAUSES
MAIN OPTION CLAUSES
SECONDARY OPTION CLAUSES
COST COMPONENTS

MAIN OPTION CLAUSES

Option A: Priced contract with activity schedule

Identified and defined terms

11

11.2 (20) The Activity Schedule is the *activity schedule* unless later changed in accordance with this contract.

(22) Defined Cost is the cost of the components in the Shorter Schedule of Cost Components whether work is subcontracted or not excluding the cost of preparing quotations for compensation events.

(27) The Price for Work Done to Date is the total of the Prices for

- each group of completed activities and
- each completed activity which is not in a group.

A completed activity is one which is without Defects which would either delay or be covered by immediately following work.

(30) The Prices are the lump sum prices for each of the activities on the Activity Schedule unless later changed in accordance with this contract.

The programme

31

31.4 The *Contractor* provides information which shows how each activity on the Activity Schedule relates to the operations on each programme which he submits for acceptance.

The Activity Schedule

54

54.1 Information in the Activity Schedule is not Works Information or Site Information.

54.2 If the *Contractor* changes a planned method of working at his discretion so that the activities on the Activity Schedule do not relate to the operations on the Accepted Programme, he submits a revision of the Activity Schedule to the *Project Manager* for acceptance.

Main Option Clauses

OPTION A: PRICED CONTRACT WITH ACTIVITY SCHEDULE

Identified and defined terms

11

11.2 (21) The Activity Schedule is the *activity schedule* unless later changed in accordance with these *conditions of contract*.

(23) Defined Cost is the cost of the components in the Short Schedule of Cost Components.

The change to this clause in Options A and B points to the revised Short Schedule of Cost Components, which in turn can use Subcontractor costs. The cost of preparing quotations for compensation events should now be included if it meets the test of clause 63.1, which is that the Defined Cost is effected by the compensation event.

(28) The People Rates are the *people rates* unless later changed in accordance with the contract.

The Short Schedule of Cost Components includes a new provision covering pre-pricing of people. This new clause states what the People Rates are.

(29) The Price for Work Done to Date is the total of the Prices for

- each group of completed activities and
- each completed activity which is not in a group.

A completed activity is one without notified Defects the correction of which will delay following work.

(32) The Prices are the lump sum prices for each of the activities on the Activity Schedule unless later changed in accordance with the contract.

The programme

31

31.4 The *Contractor* provides information which shows how each activity on the Activity Schedule relates to the operations on each programme submitted for acceptance.

The Activity Schedule

55

55.1 Information in the Activity Schedule is not Scope or Site Information. If the activities on the Activity Schedule do not relate to the Scope, the *Contractor* corrects the Activity Schedule.

The new sentence in clause 55.1 provides for where there is a difference between the Scope and the Activity Schedule for whatever reason. The Activity Schedule needs to properly reflect the Scope as it is the principal means of paying the *Contractor*.

55.3 If the *Contractor*

- changes a planned method of working at its discretion so that the activities on the Activity Schedule do not relate to the operations on the Accepted Programme or
- corrects the Activity Schedule so that the activities on the Activity Schedule relate to the Scope

the *Contractor* submits a revision of the Activity Schedule to the *Project Manager* for acceptance.

CORE CLAUSES
MAIN OPTION CLAUSES
SECONDARY OPTION CLAUSES
COST COMPONENTS

CORE CLAUSES
MAIN OPTION CLAUSES
SECONDARY OPTION CLAUSES
COST COMPONENTS

54.3 A reason for not accepting a revision of the Activity Schedule is that

- it does not comply with the Accepted Programme,
- any changed Prices are not reasonably distributed between the activities or
- the total of the Prices is changed.

Assessing compensation events

63

63.10 If the effect of a compensation event is to reduce the total Defined Cost and the event is

- a change to the Works Information or
- a correction of an assumption stated by the *Project Manager* for assessing an earlier compensation event,

the Prices are reduced.

63.12 Assessments for changed Prices for compensation events are in the form of changes to the Activity Schedule.

Implementing compensation events

65

65.4 The changes to the Prices, the Completion Date and the Key Dates are included in the notification implementing a compensation event.

Payment on termination

93

93.3 The amount due on termination is assessed without taking grouping of activities into account.

55.4 A reason for not accepting a revision of the Activity Schedule is that

- it does not relate to the operations on the Accepted Programme,
- any changed Prices are not reasonably distributed between the activities which are not completed or
- the total of the Prices is changed.

Assessing compensation events

63

63.12 If the effect of a compensation event is to reduce the total Defined Cost and the event is a change to the Scope provided by the *Client*, which the *Contractor* proposed and the *Project Manager* accepted, the Prices are reduced by an amount calculated by multiplying the assessed effect of the compensation event by the *value engineering percentage*.

This new clause introduces a value engineering incentive into Options A and B, with the share being determined by the *value engineering percentage* stated in the Contract Data. 

63.14 Assessments for changed Prices for compensation events are in the form of changes to the Activity Schedule.

63.16 If, when assessing a compensation event the People Rates do not include a rate for a category of person required, the *Project Manager* and *Contractor* may agree a new rate. If they do not agree the *Project Manager* assesses the rate based on the People Rates. The agreed or assessed rate becomes the People Rate for that category of person.

Payment on termination

93

93.3 The amount due on termination is assessed without taking grouping of activities into account.

Option B: Priced contract with bill of quantities

Identified and defined terms

11

11.2 (21) The Bill of Quantities is the *bill of quantities* as changed in accordance with this contract to accommodate implemented compensation events and for accepted quotations for acceleration.

(22) Defined Cost is the cost of the components in the Shorter Schedule of Cost Components whether work is subcontracted or not excluding the cost of preparing quotations for compensation events.

(28) The Price for Work Done to Date is the total of

- the quantity of the work which the *Contractor* has completed for each item in the Bill of Quantities multiplied by the rate and
- a proportion of each lump sum which is the proportion of the work covered by the item which the *Contractor* has completed.

Completed work is work without Defects which would either delay or be covered by immediately following work.

(31) The Prices are the lump sums and the amounts obtained by multiplying the rates by the quantities for the items in the Bill of Quantities.

The Bill of Quantities

55

55.1 Information in the Bill of Quantities is not Works Information or Site Information.

Compensation events

60

60.4 A difference between the final total quantity of work done and the quantity stated for an item in the Bill of Quantities is a compensation event if

- the difference does not result from a change to the Works Information,
- the difference causes the Defined Cost per unit of quantity to change and
- the rate in the Bill of Quantities for the item multiplied by the final total quantity of work done is more than 0.5% of the total of the Prices at the Contract Date.

If the Defined Cost per unit of quantity is reduced, the affected rate is reduced.

60.5 A difference between the final total quantity of work done and the quantity for an item stated in the Bill of Quantities which delays Completion or the meeting of the Condition stated for a Key Date is a compensation event.

60.6 The *Project Manager* corrects mistakes in the Bill of Quantities which are departures from the rules for item descriptions and for division of the work into items in the *method of measurement* or are due to ambiguities or inconsistencies. Each such correction is a compensation event which may lead to reduced Prices.

60.7 In assessing a compensation event which results from a correction of an inconsistency between the Bill of Quantities and another document, the *Contractor* is assumed to have taken the Bill of Quantities as correct.

OPTION B: PRICED CONTRACT WITH BILL OF QUANTITIES

Identified and defined terms

11

11.2 (22) The Bill of Quantities is the *bill of quantities* unless later changed in accordance with these *conditions of contract*.

(23) Defined Cost is the cost of the components in the Short Schedule of Cost Components.

(28) The People Rates are the *people rates* unless later changed in accordance with the contract.

(30) The Price for Work Done to Date is the total of

- the quantity of the work which the *Contractor* has completed for each item in the Bill of Quantities multiplied by the rate and
- a proportion of each lump sum which is the proportion of the work covered by the item which the *Contractor* has completed.

Completed work is work which is without notified Defects the correction of which will delay following work.

(33) The Prices are the lump sums and the amounts obtained by multiplying the rates by the quantities for the items in the Bill of Quantities.

The Bill of Quantities

56

56.1 Information in the Bill of Quantities is not Scope or Site Information.

Compensation events

60

60.4 A difference between the final total quantity of work done and the quantity stated for an item in the Bill of Quantities is a compensation event if

- the difference does not result from a change to the Scope,
- the difference causes the Defined Cost per unit of quantity to change and
- the rate in the Bill of Quantities for the item multiplied by the final total quantity of work done is more than 0.5% of the total of the Prices at the Contract Date.

If the Defined Cost per unit of quantity is reduced, the affected rate is reduced.

60.5 A difference between the final total quantity of work done and the quantity for an item stated in the Bill of Quantities which delays Completion or the meeting of the Condition stated for a Key Date is a compensation event.

60.6 The *Project Manager* gives an instruction to correct a mistake in the Bill of Quantities which is

- a departure from the rules for item descriptions or division of the work into items in the *method of measurement* or
- due to an ambiguity or inconsistency.

Each such correction is a compensation event which may lead to reduced Prices.

The starting point for correcting a mistake in the Bill of Quantities is now a *Project Manager's* instruction. 

60.7 In assessing a compensation event which results from a correction of an inconsistency between the Bill of Quantities and another document, the *Contractor* is assumed to have taken the Bill of Quantities as correct.

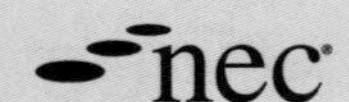

Assessing compensation events

63

63.10 If the effect of a compensation event is to reduce the total Defined Cost and the event is

- a change to the Works Information or
- a correction of an assumption stated by the *Project Manager* for assessing an earlier compensation event,

the Prices are reduced.

63.13 Assessments for changed Prices for compensation events are in the form of changes to the Bill of Quantities.

- For the whole or a part of a compensation event for work not yet done and for which there is an item in the Bill of Quantities, the changes are
  - a changed rate,
  - a changed quantity or
  - a changed lump sum.
- For the whole or a part of a compensation event for work not yet done and for which there is no item in the Bill of Quantities, the change is a new priced item which, unless the *Project Manager* and the *Contractor* agree otherwise, is compiled in accordance with the *method of measurement*.
- For the whole or a part of a compensation event for work already done, the change is a new lump sum item.

Implementing compensation events

65

65.4 The changes to the Prices, the Completion Date and the Key Dates are included in the notification implementing a compensation event.

Assessing compensation events

63

63.12 If the effect of a compensation event is to reduce the total Defined Cost and the event is a change to the Scope provided by the *Client,* which the *Contractor* proposed and the *Project Manager* accepted, the Prices are reduced by an amount calculated by multiplying the assessed value of the compensation event by the *value engineering percentage*.

63.15 Assessments for changed Prices for compensation events are in the form of changes to the Bill of Quantities.

- For the whole or a part of a compensation event for work not yet done and for which there is an item in the Bill of Quantities, the changes are
  - a changed rate,
  - a changed quantity or
  - a changed lump sum.
- For the whole or a part of a compensation event for work not yet done and for which there is no item in the Bill of Quantities, the change is a new priced item which, unless the *Project Manager* and the *Contractor* agree otherwise, is compiled in accordance with the *method of measurement.*
- For the whole or a part of a compensation event for work already done, the change is a new lump sum item.

63.16 If, when assessing a compensation event the People Rates do not include a rate for a category of person required, the *Project Manager* and *Contractor* may agree a new rate. If they do not agree the *Project Manager* assesses the rate based on the People Rates. The agreed or assessed rate becomes the People Rate for that category of person.

Option C: Target contract with activity schedule

Identified and defined terms

11

11.2 (20) The Activity Schedule is the *activity schedule* unless later changed in accordance with this contract.

(23) Defined Cost is

- the amount of payments due to Subcontractors for work which is subcontracted without taking account of amounts deducted for
  - retention,
  - payment to the *Employer* as a result of the Subcontractor failing to meet a Key Date,
  - the correction of Defects after Completion,
  - payments to Others and
  - the supply of equipment, supplies and services included in the charge for overhead cost within the Working Areas in this contract

and

- the cost of components in the Schedule of Cost Components for other work

less Disallowed Cost.

(25) Disallowed Cost is cost which the *Project Manager* decides

- is not justified by the *Contractor*'s accounts and records,
- should not have been paid to a Subcontractor or supplier in accordance with his contract,
- was incurred only because the *Contractor* did not
  - follow an acceptance or procurement procedure stated in the Works Information or
  - give an early warning which this contract required him to give

and the cost of

- correcting Defects after Completion,
- correcting Defects caused by the *Contractor* not complying with a constraint on how he is to Provide the Works stated in the Works Information,
- Plant and Materials not used to Provide the Works (after allowing for reasonable wastage) unless resulting from a change to the Works Information,
- resources not used to Provide the Works (after allowing for reasonable availability and utilisation) or not taken away from the Working Areas when the *Project Manager* requested and
- preparation for and conduct of an adjudication or proceedings of the *tribunal*.

OPTION C: TARGET CONTRACT WITH ACTIVITY SCHEDULE

Identified and defined terms

11

11.2 (21) The Activity Schedule is the *activity schedule* unless later changed in accordance with these *conditions of contract*.

(24) Defined Cost is the cost of the components in the Schedule of Cost Components less Disallowed Cost.

A shorter definition of Defined Cost is now provided for Options C to E. It is now principally the rules in the Schedule of Cost Components that determine how Defined Cost is calculated.

(26) Disallowed Cost is cost which

- is not justified by the *Contractor's* accounts and records,
- should not have been paid to a Subcontractor or supplier in accordance with its contract,
- was incurred only because the *Contractor* did not
  - follow an acceptance or procurement procedure stated in the Scope,
  - give an early warning which the contract required it to give or
  - give notification to the *Project Manager* of the preparation for and conduct of an adjudication or proceedings of a tribunal between the *Contractor* and a Subcontractor or supplier

and the cost of

- correcting Defects after Completion,
- correcting Defects caused by the *Contractor* not complying with a constraint on how it is to Provide the Works stated in the Scope,
- Plant and Materials not used to Provide the Works (after allowing for reasonable wastage) unless resulting from a change to the Scope,
- resources not used to Provide the Works (after allowing for reasonable availability and utilisation) or not taken away from the Working Areas when the *Project Manager* requested and
- preparation for and conduct of an adjudication, payments to a member of the Dispute Avoidance Board or proceedings of the *tribunal* between the Parties.

A new Disallowed Cost has been added to ensure the *Contractor* notifies an early warning in the event that a dispute the *Contractor* has with a Subcontractor or supplier is likely to lead to an adjudication or a tribunal. This gives the *Project Manager* an opportunity to attempt to resolve the issue on behalf of the *Client*.

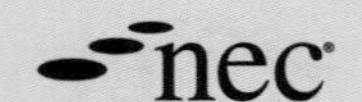

(29) The Price for Work Done to Date is the total Defined Cost which the *Project Manager* forecasts will have been paid by the *Contractor* before the next assessment date plus the Fee.

(30) The Prices are the lump sum prices for each of the activities on the Activity Schedule unless later changed in accordance with this contract.

Providing the Works **20**

20.3 The *Contractor* advises the *Project Manager* on the practical implications of the design of the *works* and on subcontracting arrangements.

20.4 The *Contractor* prepares forecasts of the total Defined Cost for the whole of the *works* in consultation with the *Project Manager* and submits them to the *Project Manager*. Forecasts are prepared at the intervals stated in the Contract Data from the *starting date* until Completion of the whole of the *works*. An explanation of the changes made since the previous forecast is submitted with each forecast.

Subcontracting **26**

26.4 The *Contractor* submits the proposed contract data for each subcontract for acceptance to the *Project Manager* if

- an NEC contract is proposed and
- the *Project Manager* instructs the *Contractor* to make the submission.

A reason for not accepting the proposed contract data is that its use will not allow the *Contractor* to Provide the Works.

The programme **31**

31.4 The *Contractor* provides information which shows how each activity on the Activity Schedule relates to the operations on each programme which he submits for acceptance.

Tests and inspections **40**

40.7 When the *Project Manager* assesses the cost incurred by the *Employer* in repeating a test or inspection after a Defect is found, the *Project Manager* does not include the *Contractor*'s cost of carrying out the repeat test or inspection.

Assessing the amount due **50**

50.6 Payments of Defined Cost made by the *Contractor* in a currency other than the *currency of this contract* are included in the amount due as payments to be made to him in the same currency. Such payments are converted to the *currency of this contract* in order to calculate the Fee and any *Contractor*'s share using the *exchange rates.*

(31) The Price for Work Done to Date is the total Defined Cost which the *Project Manager* forecasts will have been paid by the *Contractor* before the next assessment date plus the Fee.

(32) The Prices are the lump sum prices for each of the activities on the Activity Schedule unless later changed in accordance with the contract.

Providing the Works 20

20.3 The *Contractor* advises the *Project Manager* on the practical implications of the design of the *works* and on subcontracting arrangements.

20.4 The *Contractor* prepares forecasts of the total Defined Cost for the whole of the *works* in consultation with the *Project Manager* and submits them to the *Project Manager*. Forecasts are prepared at the intervals stated in the Contract Data from the *starting date* until Completion of the whole of the *works*. An explanation of the changes made since the previous forecast is submitted with each forecast.

Subcontracting 26

26.4 The *Contractor* submits the pricing information in the proposed subcontract documents for each subcontract to the *Project Manager* unless the *Project Manager* has agreed that no submission is required.

Tests and inspections 41

41.7 When the *Project Manager* assesses the cost incurred by the *Client* in repeating a test or inspection after a Defect is found, the *Project Manager* does not include the *Contractor's* cost of carrying out the repeat test or inspection.

Assessing the amount due 50

50.7 Payments of Defined Cost made by the *Contractor* in a currency other than the *currency of the contract* are included in the amount due as payments to be made to it in the same currency. Such payments are converted to the *currency of the contract* in order to calculate the Fee and any *Contractor's* share using the *exchange rates*.

50.9 The *Contractor* notifies the *Project Manager* when a part of Defined Cost has been finalised, and makes available for inspection the records necessary to demonstrate that it has been correctly assessed. The *Project Manager* reviews the records made available, and no later than thirteen weeks after the *Contractor's* notification

- accepts that part of Defined Cost as correct,
- notifies the *Contractor* that further records are needed or
- notifies the *Contractor* of errors in its assessment.

The *Contractor* provides any further records requested or advises the correction of the errors in its assessment within four weeks of the *Project Manager's* notification. The *Project Manager* reviews the records provided, and within four weeks

- accepts the cost as correct or
- notifies the *Contractor* of the correct assessment of that part of Defined Cost.

If the *Project Manager* does not notify a decision on that part of Defined Cost within the time stated, the *Contractor's* assessment is treated as correct.

This new process added into Options C to F allows the *Contractor* to finalise parts of the Defined Cost, rather than leaving them open until the defects date. This should reduce the number of problems with Defined Costs that only arise at a very late stage.

Defined Cost

52

52.2 The *Contractor* keeps these records

- accounts of payments of Defined Cost,
- proof that the payments have been made,
- communications about and assessments of compensation events for Subcontractors and
- other records as stated in the Works Information.

52.3 The *Contractor* allows the *Project Manager* to inspect at any time within working hours the accounts and records which he is required to keep.

The *Contractor*'s share

53

53.1 The *Project Manager* assesses the *Contractor*'s share of the difference between the total of the Prices and the Price for Work Done to Date. The difference is divided into increments falling within each of the *share ranges*. The limits of a *share range* are the Price for Work Done to Date divided by the total of the Prices, expressed as a percentage. The *Contractor*'s share equals the sum of the products of the increment within each *share range* and the corresponding *Contractor's share percentage.*

53.2 If the Price for Work Done to Date is less than the total of the Prices, the *Contractor* is paid his share of the saving. If the Price for Work Done to Date is greater than the total of the Prices, the *Contractor* pays his share of the excess.

53.3 The *Project Manager* makes a preliminary assessment of the *Contractor*'s share at Completion of the whole of the *works* using his forecasts of the final Price for Work Done to Date and the final total of the Prices. This share is included in the amount due following Completion of the whole of the *works*.

53.4 The *Project Manager* makes a final assessment of the *Contractor*'s share using the final Price for Work Done to Date and the final total of the Prices. This share is included in the final amount due.

The Activity Schedule

54

54.1 Information in the Activity Schedule is not Works Information or Site Information.

54.2 If the *Contractor* changes a planned method of working at his discretion so that the activities on the Activity Schedule do not relate to the operations on the Accepted Programme, he submits a revision of the Activity Schedule to the *Project Manager* for acceptance.

54.3 A reason for not accepting a revision of the Activity Schedule is that

- it does not comply with the Accepted Programme,
- any changed Prices are not reasonably distributed between the activities or
- the total of the Prices is changed.

Defined Cost 52

52.2 The *Contractor* keeps these records

- accounts of payments of Defined Cost,
- proof that the payments have been made,
- communications about and assessments of compensation events for Subcontractors and
- other records as stated in the Scope.

52.4 The *Contractor* allows the *Project Manager* to inspect at any time within working hours the accounts and records which it is required to keep.

The *Contractor's* share 54

54.1 The *Project Manager* assesses the *Contractor's* share of the difference between the total of the Prices and the Price for Work Done to Date. The difference is divided into increments falling within each of the *share ranges*. The limits of a *share range* are the Price for Work Done to Date divided by the total of the Prices, expressed as a percentage. The *Contractor's* share equals the sum of the products of the increment within each *share range* and the corresponding *Contractor's share percentage*.

54.2 If the Price for Work Done to Date is less than the total of the Prices, the *Contractor* is paid its share of the saving. If the Price for Work Done to Date is greater than the total of the Prices, the *Contractor* pays its share of the excess.

54.3 The *Project Manager* makes a preliminary assessment of the *Contractor's* share at Completion of the whole of the *works* using forecasts of the final Price for Work Done to Date and the final total of the Prices. This share is included in the amount due following Completion of the whole of the *works*.

54.4 The *Project Manager* makes a final assessment of the *Contractor's* share using the final Price for Work Done to Date and the final total of the Prices. This share is included in the final amount due.

The Activity Schedule 55

55.2 Information in the Activity Schedule is not Scope or Site Information.

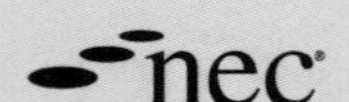

Assessing compensation events

63

63.11 If the effect of a compensation event is to reduce the total Defined Cost and the event is

- a change to the Works Information, other than a change to the Works Information provided by the *Employer* which the *Contractor* proposed and the *Project Manager* has accepted or
- a correction of an assumption stated by the *Project Manager* for assessing an earlier compensation event,

the Prices are reduced.

63.12 Assessments for changed Prices for compensation events are in the form of changes to the Activity Schedule.

63.15 If the *Project Manager* and the *Contractor* agree, the *Contractor* assesses a compensation event using the Shorter Schedule of Cost Components. The *Project Manager* may make his own assessments using the Shorter Schedule of Cost Components.

Implementing compensation events

65

65.4 The changes to the Prices, the Completion Date and the Key Dates are included in the notification implementing a compensation event.

Payment on termination

93

93.4 If there is a termination, the *Project Manager* assesses the *Contractor*'s share after he has certified termination. His assessment uses, as the Price for Work Done to Date, the total of the Defined Cost which the *Contractor* has paid and which he is committed to pay for work done before termination. The assessment uses as the total of the Prices

- the lump sum price for each activity which has been completed and
- a proportion of the lump sum price for each incomplete activity which is the proportion of the work in the activity which has been completed.

93.6 The *Project Manager*'s assessment of the *Contractor*'s share is added to the amount due to the *Contractor* on termination if there has been a saving or deducted if there has been an excess.

Assessing compensation events

63

63.13 If the effect of a compensation event is to reduce the total Defined Cost and the event is a change to the Scope provided by the *Client,* which the *Contractor* proposed and the *Project Manager* accepted, the Prices are not reduced.

> Although the wording is changed here, the principle is the same. A value engineering-type proposal put forward by the *Contractor* in Option C or Option D is dealt with not by changing the target, but instead by the Parties sharing any saving through the *Contractor's* share mechanism. 

63.14 Assessments for changed Prices for compensation events are in the form of changes to the Activity Schedule.

Payment on termination

93

93.4 If there is a termination, the *Project Manager* assesses the *Contractor's* share after certifying termination. The assessment uses as the Price for Work Done to Date the total of the Defined Cost which the *Contractor* has paid and which it is committed to pay for work done before termination, and uses as the total of the Prices

- the lump sum price for each activity which has been completed and
- a proportion of the lump sum price for each incomplete activity which is the proportion of the work in the activity which has been completed.

93.6 The *Project Manager's* assessment of the *Contractor's* share is added to the amount due to the *Contractor* on termination if there has been a saving or deducted if there has been an excess.

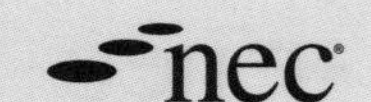

Option D: Target contract with bill of quantities

Identified and defined terms

11

11.2 (21) The Bill of Quantities is the *bill of quantities* as changed in accordance with this contract to accommodate implemented compensation events and for accepted quotations for acceleration.

(23) Defined Cost is

- the amount of payments due to Subcontractors for work which is subcontracted without taking account of amounts deducted for
  - retention,
  - payment to the *Employer* as a result of the Subcontractor failing to meet a Key Date,
  - the correction of Defects after Completion,
  - payments to Others and
  - the supply of equipment, supplies and services included in the charge for overhead cost within the Working Areas in this contract

and

- the cost of components in the Schedule of Cost Components for other work

less Disallowed Cost.

(25) Disallowed Cost is cost which the *Project Manager* decides

- is not justified by the *Contractor*'s accounts and records,
- should not have been paid to a Subcontractor or supplier in accordance with his contract,
- was incurred only because the *Contractor* did not
  - follow an acceptance or procurement procedure stated in the Works Information or
  - give an early warning which this contract required him to give

and the cost of

- correcting Defects after Completion,
- correcting Defects caused by the *Contractor* not complying with a constraint on how he is to Provide the Works stated in the Works Information,
- Plant and Materials not used to Provide the Works (after allowing for reasonable wastage) unless resulting from a change to the Works Information,
- resources not used to Provide the Works (after allowing for reasonable availability and utilisation) or not taken away from the Working Areas when the *Project Manager* requested and
- preparation for and conduct of an adjudication or proceedings of the *tribunal*.

OPTION D: TARGET CONTRACT WITH BILL OF QUANTITIES

Identified and defined terms 11

11.2 (22) The Bill of Quantities is the *bill of quantities* unless later changed in accordance with these *conditions of contract*.

(24) Defined Cost is the cost of the components in the Schedule of Cost Components less Disallowed Cost.

(26) Disallowed Cost is cost which

- is not justified by the *Contractor's* accounts and records,
- should not have been paid to a Subcontractor or supplier in accordance with its contract,
- was incurred only because the *Contractor* did not
  - follow an acceptance or procurement procedure stated in the Scope,
  - give an early warning which the contract required it to give or
  - give notification to the *Project Manager* of the preparation for and conduct of an adjudication or proceedings of a tribunal between the *Contractor* and a Subcontractor or supplier

and the cost of

- correcting Defects after Completion,
- correcting Defects caused by the *Contractor* not complying with a constraint on how it is to Provide the Works stated in the Scope,
- Plant and Materials not used to Provide the Works (after allowing for reasonable wastage) unless resulting from a change to the Scope,
- resources not used to Provide the Works (after allowing for reasonable availability and utilisation) or not taken away from the Working Areas when the *Project Manager* requested and
- preparation for and conduct of an adjudication, or payments to a member of the Dispute Avoidance Board or proceedings of the *tribunal* between the Parties.

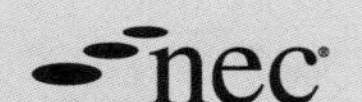

(29) The Price for Work Done to Date is the total Defined Cost which the *Project Manager* forecasts will have been paid by the *Contractor* before the next assessment date plus the Fee.

(31) The Prices are the lump sums and the amounts obtained by multiplying the rates by the quantities for the items in the Bill of Quantities.

(33) The Total of the Prices is the total of

- the quantity of the work which the *Contractor* has completed for each item in the Bill of Quantities multiplied by the rate and
- a proportion of each lump sum which is the proportion of the work covered by the item which the *Contractor* has completed.

Completed work is work without Defects which would either delay or be covered by immediately following work.

Providing the Works **20**

20.3 The *Contractor* advises the *Project Manager* on the practical implications of the design of the *works* and on subcontracting arrangements.

20.4 The *Contractor* prepares forecasts of the total Defined Cost for the whole of the *works* in consultation with the *Project Manager* and submits them to the *Project Manager*. Forecasts are prepared at the intervals stated in the Contract Data from the *starting date* until Completion of the whole of the *works*. An explanation of the changes made since the previous forecast is submitted with each forecast.

Subcontracting **26**

26.4 The *Contractor* submits the proposed contract data for each subcontract for acceptance to the *Project Manager* if

- an NEC contract is proposed and
- the *Project Manager* instructs the *Contractor* to make the submission.

A reason for not accepting the proposed contract data is that its use will not allow the *Contractor* to Provide the Works.

Tests and inspections **40**

40.7 When the *Project Manager* assesses the cost incurred by the *Employer* in repeating a test or inspection after a Defect is found, the *Project Manager* does not include the *Contractor*'s cost of carrying out the repeat test or inspection.

Assessing the amount due **50**

50.6 Payments of Defined Cost made by the *Contractor* in a currency other than the *currency of this contract* are included in the amount due as payments to be made to him in the same currency. Such payments are converted to the *currency of this contract* in order to calculate the Fee and any *Contractor*'s share using the *exchange rates*.

(31) The Price for Work Done to Date is the total Defined Cost which the *Project Manager* forecasts will have been paid by the *Contractor* before the next assessment date plus the Fee.

(33) The Prices are the lump sums and the amounts obtained by multiplying the rates by the quantities for the items in the Bill of Quantities.

(35) The Total of the Prices is the total of

- the quantity of the work which the *Contractor* has completed for each item in the Bill of Quantities multiplied by the rate and
- a proportion of each lump sum which is the proportion of the work covered by the item which the *Contractor* has completed.

Completed work is work which is without notified Defects the correction of which will delay following work.

Providing the Works

20

20.3 The *Contractor* advises the *Project Manager* on the practical implications of the design of the *works* and on subcontracting arrangements.

20.4 The *Contractor* prepares forecasts of the total Defined Cost for the whole of the *works* in consultation with the *Project Manager* and submits them to the *Project Manager*. Forecasts are prepared at the intervals stated in the Contract Data from the *starting date* until Completion of the whole of the *works*. An explanation of the changes made since the previous forecast is submitted with each forecast.

Subcontracting

26

26.4 The *Contractor* submits the pricing information in the proposed subcontract documents for each subcontract to the *Project Manager* unless the *Project Manager* has agreed that no submission is required.

Tests and inspections

41

41.7 When the *Project Manager* assesses the cost incurred by the *Client* in repeating a test or inspection after a Defect is found, the *Project Manager* does not include the *Contractor's* cost of carrying out the repeat test or inspection.

Assessing the amount due

50

50.7 Payments of Defined Cost made by the *Contractor* in a currency other than the *currency of the contract* are included in the amount due as payments to be made to it in the same currency. Such payments are converted to the *currency of the contract* in order to calculate the Fee and any *Contractor's* share using the *exchange rates*.

50.9 The *Contractor* notifies the *Project Manager* when a part of Defined Cost has been finalised, and makes available for inspection the records necessary to demonstrate that it has been correctly assessed. The *Project Manager* reviews the records made available, and no later than thirteen weeks after the *Contractor's* notification

- accepts that part of Defined Cost as correct,
- notifies the *Contractor* that further records are needed or
- notifies the *Contractor* of errors in its assessment.

The *Contractor* provides any further records requested or advises the correction of the errors in its assessment within four weeks of the *Project Manager's* notification. The *Project Manager* reviews the records provided, and within four weeks

- accepts the cost as correct or
- notifies the *Contractor* of the correct assessment of that part of Defined Cost.

If the *Project Manager* does not notify a decision on that part of Defined Cost within the time stated, the *Contractor's* assessment is treated as correct.

Defined Cost

52

52.2 The *Contractor* keeps these records

- accounts of payments of Defined Cost,
- proof that the payments have been made,
- communications about and assessments of compensation events for Subcontractors and
- other records as stated in the Works Information.

52.3 The *Contractor* allows the *Project Manager* to inspect at any time within working hours the accounts and records which he is required to keep.

The *Contractor*'s share

53

53.5 The *Project Manager* assesses the *Contractor*'s share of the difference between the Total of the Prices and the Price for Work Done to Date. The difference is divided into increments falling within each of the *share ranges*. The limits of a *share range* are the Price for Work Done to Date divided by the Total of the Prices, expressed as a percentage. The *Contractor*'s share equals the sum of the products of the increment within each *share range* and the corresponding *Contractor's share percentage*.

53.6 If the Price for Work Done to Date is less than the Total of the Prices, the *Contractor* is paid his share of the saving. If the Price for Work Done to Date is greater than the Total of the Prices, the *Contractor* pays his share of the excess.

53.7 The *Project Manager* makes a preliminary assessment of the *Contractor*'s share at Completion of the whole of the *works* using his forecasts of the final Price for Work Done to Date and the final Total of the Prices. This share is included in the amount due following Completion of the whole of the *works*.

53.8 The *Project Manager* makes a final assessment of the *Contractor*'s share using the final Price for Work Done to Date and the final Total of the Prices. This share is included in the final amount due.

The Bill of Quantities

55

55.1 Information in the Bill of Quantities is not Works Information or Site Information.

Compensation events

60

60.4 A difference between the final total quantity of work done and the quantity stated for an item in the Bill of Quantities is a compensation event if

- the difference does not result from a change to the Works Information,
- the difference causes the Defined Cost per unit of quantity to change and
- the rate in the Bill of Quantities for the item multiplied by the final total quantity of work done is more than 0.5% of the total of the Prices at the Contract Date.

If the Defined Cost per unit of quantity is reduced, the affected rate is reduced.

60.5 A difference between the final total quantity of work done and the quantity for an item stated in the Bill of Quantities which delays Completion or the meeting of the Condition stated for a Key Date is a compensation event.

Defined Cost	**52**	
	52.2	The *Contractor* keeps these records • accounts of payments of Defined Cost, • proof that the payments have been made, • communications about and assessments of compensation events for Subcontractors and • other records as stated in the Scope.
	52.4	The *Contractor* allows the *Project Manager* to inspect at any time within working hours the accounts and records which it is required to keep.
The *Contractor's* share	**54**	
	54.5	The *Project Manager* assesses the *Contractor's* share of the difference between the Total of the Prices and the Price for Work Done to Date. The difference is divided into increments falling within each of the *share ranges*. The limits of a *share range* are the Price for Work Done to Date divided by the Total of the Prices, expressed as a percentage. The *Contractor's* share equals the sum of the products of the increment within each *share range* and the corresponding *Contractor's share percentage*.
	54.6	If the Price for Work Done to Date is less than the Total of the Prices, the *Contractor* is paid its share of the saving. If the Price for Work Done to Date is greater than the Total of the Prices, the *Contractor* pays its share of the excess.
	54.7	The *Project Manager* makes a preliminary assessment of the *Contractor's* share at Completion of the whole of the *works* using forecasts of the final Price for Work Done to Date and the final Total of the Prices. This share is included in the amount due following Completion of the whole of the *works*.
	54.8	The *Project Manager* makes a final assessment of the *Contractor's* share using the final Price for Work Done to Date and the final Total of the Prices. This share is included in the final amount due.
The Bill of Quantities	**56**	
	56.1	Information in the Bill of Quantities is not Scope or Site Information.
Compensation events	**60**	
	60.4	A difference between the final total quantity of work done and the quantity stated for an item in the Bill of Quantities is a compensation event if • the difference does not result from a change to the Scope, • the difference causes the Defined Cost per unit of quantity to change and • the rate in the Bill of Quantities for the item multiplied by the final total quantity of work done is more than 0.5% of the total of the Prices at the Contract Date. If the Defined Cost per unit of quantity is reduced, the affected rate is reduced.
	60.5	A difference between the final total quantity of work done and the quantity for an item stated in the Bill of Quantities which delays Completion or the meeting of the Condition stated for a Key Date is a compensation event.

60.6 The *Project Manager* corrects mistakes in the Bill of Quantities which are departures from the rules for item descriptions and for division of the work into items in the *method of measurement* or are due to ambiguities or inconsistencies. Each such correction is a compensation event which may lead to reduced Prices.

60.7 In assessing a compensation event which results from a correction of an inconsistency between the Bill of Quantities and another document, the *Contractor* is assumed to have taken the Bill of Quantities as correct.

Assessing compensation events **63**

63.11 If the effect of a compensation event is to reduce the total Defined Cost and the event is

- a change to the Works Information, other than a change to the Works Information provided by the *Employer* which the *Contractor* proposed and the *Project Manager* has accepted or
- a correction of an assumption stated by the *Project Manager* for assessing an earlier compensation event,

the Prices are reduced.

63.13 Assessments for changed Prices for compensation events are in the form of changes to the Bill of Quantities.

- For the whole or a part of a compensation event for work not yet done and for which there is an item in the Bill of Quantities, the changes are
  - a changed rate,
  - a changed quantity or
  - a changed lump sum.
- For the whole or a part of a compensation event for work not yet done and for which there is no item in the Bill of Quantities, the change is a new priced item which, unless the *Project Manager* and the *Contractor* agree otherwise, is compiled in accordance with the *method of measurement.*
- For the whole or a part of a compensation event for work already done, the change is a new lump sum item.

63.15 If the *Project Manager* and the *Contractor* agree, the *Contractor* assesses a compensation event using the Shorter Schedule of Cost Components. The *Project Manager* may make his own assessments using the Shorter Schedule of Cost Components.

Implementing compensation events **65**

65.4 The changes to the Prices, the Completion Date and the Key Dates are included in the notification implementing a compensation event.

Payment on termination **93**

93.5 If there is a termination, the *Project Manager* assesses the *Contractor*'s share after he has certified termination. His assessment uses, as the Price for Work Done to Date, the total of the Defined Cost which the *Contractor* has paid and which he is committed to pay for work done before termination.

93.6 The *Project Manager*'s assessment of the *Contractor*'s share is added to the amounts due to the *Contractor* on termination if there has been a saving or deducted if there has been an excess.

60.6 The *Project Manager* gives an instruction to correct a mistake in the Bill of Quantities which is

- a departure from the rules for item descriptions or division of the work into items in the *method of measurement* or
- due to an ambiguity or inconsistency.

Each such correction is a compensation event which may lead to reduced Prices.

60.7 In assessing a compensation event which results from a correction of an inconsistency between the Bill of Quantities and another document, the *Contractor* is assumed to have taken the Bill of Quantities as correct.

Assessing compensation events

63

63.13 If the effect of a compensation event is to reduce the total Defined Cost and the event is a change to the Scope provided by the *Client*, which the *Contractor* proposed and the *Project Manager* accepted, the prices are not reduced.

63.15 Assessments for changed Prices for compensation events are in the form of changes to the Bill of Quantities.

- For the whole or a part of a compensation event for work not yet done and for which there is an item in the Bill of Quantities, the changes are
  - a changed rate,
  - a changed quantity or
  - a changed lump sum.
- For the whole or a part of a compensation event for work not yet done and for which there is no item in the Bill of Quantities, the change is a new priced item which, unless the *Project Manager* and the *Contractor* agree otherwise, is compiled in accordance with the *method of measurement*.
- For the whole or a part of a compensation event for work already done, the change is a new lump sum item.

Payment on termination

93

93.5 If there is a termination, the *Project Manager* assesses the *Contractor's* share after certifying termination. The assessment uses, as the Price for Work Done to Date, the total of the Defined Cost which the *Contractor* has paid and which it is committed to pay for work done before termination.

93.6 The *Project Manager's* assessment of the *Contractor's* share is added to the amounts due to the *Contractor* on termination if there has been a saving or deducted if there has been an excess.

CORE CLAUSES

MAIN OPTION CLAUSES

SECONDARY OPTION CLAUSES

COST COMPONENTS

Option E: Cost reimbursable contract

Identified and defined terms

11

11.2 (23) Defined Cost is

- the amount of payments due to Subcontractors for work which is subcontracted without taking account of amounts deducted for
  - retention,
  - payment to the *Employer* as a result of the Subcontractor failing to meet a Key Date,
  - the correction of Defects after Completion,
  - payments to Others and
  - the supply of equipment, supplies and services included in the charge for overhead cost within the Working Areas in this contract

and

- the cost of components in the Schedule of Cost Components for other work

less Disallowed Cost.

(25) Disallowed Cost is cost which the *Project Manager* decides

- is not justified by the *Contractor*'s accounts and records,
- should not have been paid to a Subcontractor or supplier in accordance with his contract,
- was incurred only because the *Contractor* did not
  - follow an acceptance or procurement procedure stated in the Works Information or
  - give an early warning which this contract required him to give

and the cost of

- correcting Defects after Completion,
- correcting Defects caused by the *Contractor* not complying with a constraint on how he is to Provide the Works stated in the Works Information,
- Plant and Materials not used to Provide the Works (after allowing for reasonable wastage) unless resulting from a change to the Works Information,
- resources not used to Provide the Works (after allowing for reasonable availability and utilisation) or not taken away from the Working Areas when the *Project Manager* requested and
- preparation for and conduct of an adjudication or proceedings of the *tribunal*.

(29) The Price for Work Done to Date is the total Defined Cost which the *Project Manager* forecasts will have been paid by the *Contractor* before the next assessment date plus the Fee.

(32) The Prices are the Defined Cost plus the Fee.

OPTION E: COST REIMBURSABLE CONTRACT

Identified and defined terms

11

11.2 (24) Defined Cost is the cost of the components in the Schedule of Cost Components less Disallowed Cost.

(26) Disallowed Cost is cost which

- is not justified by the *Contractor's* accounts and records,
- should not have been paid to a Subcontractor or supplier in accordance with its contract,
- was incurred only because the *Contractor* did not
  - follow an acceptance or procurement procedure stated in the Scope,
  - give an early warning which the contract required it to give or
  - give notification to the *Project Manager* of the preparation for and conduct of an adjudication or proceedings of a tribunal between the *Contractor* and a Subcontractor or supplier

and the cost of

- correcting Defects after Completion,
- correcting Defects caused by the *Contractor* not complying with a constraint on how it is to Provide the Works stated in the Scope,
- Plant and Materials not used to Provide the Works (after allowing for reasonable wastage) unless resulting from a change to the Scope,
- resources not used to Provide the Works (after allowing for reasonable availability and utilisation) or not taken away from the Working Areas when the *Project Manager* requested and
- preparation for and conduct of an adjudication, or payments to a member of the Dispute Avoidance Board or proceedings of the *tribunal* between the Parties.

(31) The Price for Work Done to Date is the total Defined Cost which the *Project Manager* forecasts will have been paid by the *Contractor* before the next assessment date plus the Fee.

(34) The Prices are the forecast of the total Defined Cost for the whole of the *works* plus the Fee.

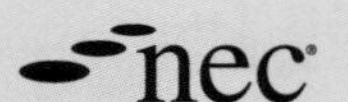

Providing the Works **20**

20.3 The *Contractor* advises the *Project Manager* on the practical implications of the design of the *works* and on subcontracting arrangements.

20.4 The *Contractor* prepares forecasts of the total Defined Cost for the whole of the *works* in consultation with the *Project Manager* and submits them to the *Project Manager*. Forecasts are prepared at the intervals stated in the Contract Data from the *starting date* until Completion of the whole of the *works*. An explanation of the changes made since the previous forecast is submitted with each forecast.

Subcontracting **26**

26.4 The *Contractor* submits the proposed contract data for each subcontract for acceptance to the *Project Manager* if

- an NEC contract is proposed and
- the *Project Manager* instructs the *Contractor* to make the submission.

A reason for not accepting the proposed contract data is that its use will not allow the *Contractor* to Provide the Works.

Tests and inspections **40**

40.7 When the *Project Manager* assesses the cost incurred by the *Employer* in repeating a test or inspection after a Defect is found, the *Project Manager* does not include the *Contractor's* cost of carrying out the repeat test or inspection.

Assessing the amount due **50**

50.7 Payments of Defined Cost made by the *Contractor* in a currency other than the *currency of this contract* are included in the amount due as payments to be made to him in the same currency. Such payments are converted to the *currency of this contract* in order to calculate the Fee using the *exchange rates*.

Defined Cost **52**

52.2 The *Contractor* keeps these records

- accounts of payments of Defined Cost,
- proof that the payments have been made,
- communications about and assessments of compensation events for Subcontractors and
- other records as stated in the Works Information.

52.3 The *Contractor* allows the *Project Manager* to inspect at any time within working hours the accounts and records which he is required to keep.

Assessing compensation events **63**

63.15 If the *Project Manager* and the *Contractor* agree, the *Contractor* assesses a compensation event using the Shorter Schedule of Cost Components. The *Project Manager* may make his own assessments using the Shorter Schedule of Cost Components.

Implementing compensation events **65**

65.3 The changes to the forecast amount of the Prices, the Completion Date and the Key Dates are included in the notification implementing a compensation event.

Providing the Works 20

20.3 The *Contractor* advises the *Project Manager* on the practical implications of the design of the *works* and on subcontracting arrangements.

20.4 The *Contractor* prepares forecasts of the total Defined Cost for the whole of the *works* in consultation with the *Project Manager* and submits them to the *Project Manager*. Forecasts are prepared at the intervals stated in the Contract Data from the *starting date* until Completion of the whole of the *works*. An explanation of the changes made since the previous forecast is submitted with each forecast.

Subcontracting 26

26.4 The *Contractor* submits the pricing information in the proposed subcontract documents for each subcontract to the *Project Manager* unless the *Project Manager* has agreed that no submission is required.

Tests and inspections 41

41.7 When the *Project Manager* assesses the cost incurred by the *Client* in repeating a test or inspection after a Defect is found, the *Project Manager* does not include the *Contractor's* cost of carrying out the repeat test or inspection.

Assessing the amount due 50

50.8 Payments of Defined Cost made by the *Contractor* in a currency other than the *currency of the contract* are included in the amount due as payments to be made to it in the same currency. Such payments are converted to the *currency of the contract* in order to calculate the Fee using the *exchange rates*.

50.9 The *Contractor* notifies the *Project Manager* when a part of Defined Cost has been finalised, and makes available for inspection the records necessary to demonstrate that it has been correctly assessed. The *Project Manager* reviews the records made available, and no later than thirteen weeks after the *Contractor's* notification

- accepts that part of Defined Cost as correct,
- notifies the *Contractor* that further records are needed or
- notifies the *Contractor* of errors in its assessment.

The *Contractor* provides any further records requested or advises the correction of the errors in its assessment within four weeks of the *Project Manager's* notification. The *Project Manager* reviews the records provided, and within four weeks

- accepts the cost as correct or
- notifies the *Contractor* of the correct assessment of that part of Defined Cost.

If the *Project Manager* does not notify a decision on that part of Defined Cost within the time stated, the *Contractor's* assessment is treated as correct.

Defined Cost 52

52.2 The *Contractor* keeps these records

- accounts of payments of Defined Cost,
- proof that the payments have been made,
- communications about and assessments of compensation events for Subcontractors and
- other records as stated in the Scope.

52.4 The *Contractor* allows the *Project Manager* to inspect at any time within working hours the accounts and records which it is required to keep.

Option F: Management contract

Identified and defined terms

11

11.2 (24) Defined Cost is

- the amount of payments due to Subcontractors for work which is subcontracted without taking account of amounts deducted for
  - retention,
  - payment to the *Employer* as a result of the Subcontractor failing to meet a Key Date,
  - the correction of Defects after Completion,
  - payments to Others,
  - the supply of equipment, supplies and services included in the charge for overhead cost within the Working Areas in this contract

and

- the *prices* for work done by the *Contractor* himself

less Disallowed Cost.

(26) Disallowed Cost is cost which the *Project Manager* decides

- is not justified by the accounts and records provided by the *Contractor*,
- should not have been paid to a Subcontractor or supplier in accordance with his contract,
- was incurred only because the *Contractor* did not
  - follow an acceptance or procurement procedure stated in the Works Information or
  - give an early warning which this contract required him to give or
- is a payment to a Subcontractor for
  - work which the Contract Data states that the *Contractor* will do himself or
  - the *Contractor*'s management.

(29) The Price for Work Done to Date is the total Defined Cost which the *Project Manager* forecasts will have been paid by the *Contractor* before the next assessment date plus the Fee.

(32) The Prices are the Defined Cost plus the Fee.

OPTION F: MANAGEMENT CONTRACT

Identified and defined terms

11

11.2 (25) Defined Cost is

- the amount of payments due to Subcontractors for work which is subcontracted without taking account of amounts paid to or retained from the Subcontractor by the *Contractor* which would result in the *Client* paying or retaining the amount twice and
- the *prices* for work done by the *Contractor*

less Disallowed Cost.

(27) Disallowed Cost is cost which

- is not justified by the *Contractor's* accounts and records,
- should not have been paid to a Subcontractor or supplier in accordance with its contract,
- was incurred only because the *Contractor* did not
  - follow an acceptance or procurement procedure stated in the Scope,
  - give an early warning which the contract required it to give or
  - give notification to the *Project Manager* of the preparation for and conduct of an adjudication or proceedings of a tribunal between the *Contractor* and a Subcontractor or
- is a payment to a Subcontractor for
  - work which the Contract Data states that the *Contractor* will do themselves or
  - the *Contractor's* management
- and was incurred in the preparation for and conduct of an adjudication, or payments to a member of the Dispute Avoidance Board or proceedings of the *tribunal* between the Parties.

(31) The Price for Work Done to Date is the total Defined Cost which the *Project Manager* forecasts will have been paid by the *Contractor* before the next assessment date plus the Fee.

(34) The Prices are the forecast of the total Defined Cost for the whole of the *works* plus the Fee.

Providing the Works **20**

20.2 The *Contractor* manages the *Contractor*'s design, the provision of Site services and the construction and installation of the *works*. The *Contractor* subcontracts the *Contractor*'s design, the provision of Site services and the construction and installation of the *works* except work which the Contract Data states that he will do himself.

20.3 The *Contractor* advises the *Project Manager* on the practical implications of the design of the *works* and on subcontracting arrangements.

20.4 The *Contractor* prepares forecasts of the total Defined Cost for the whole of the *works* in consultation with the *Project Manager* and submits them to the *Project Manager*. Forecasts are prepared at the intervals stated in the Contract Data from the *starting date* until Completion of the whole of the *works*. An explanation of the changes made since the previous forecast is submitted with each forecast.

20.5 If work which the *Contractor* is to do himself is affected by a compensation event, the *Project Manager* and the *Contractor* agree the change to the price for the work and any change to the Completion Date and Key Dates. If they cannot agree, the *Project Manager* decides the change.

Subcontracting **26**

26.4 The *Contractor* submits the proposed contract data for each subcontract for acceptance to the *Project Manager* if

- an NEC contract is proposed and
- the *Project Manager* instructs the *Contractor* to make the submission.

A reason for not accepting the proposed contract data is that its use will not allow the *Contractor* to Provide the Works.

Assessing the amount due **50**

50.7 Payments of Defined Cost made by the *Contractor* in a currency other than the *currency of this contract* are included in the amount due as payments to be made to him in the same currency. Such payments are converted to the *currency of this contract* in order to calculate the Fee using the *exchange rates*.

Defined Cost **52**

52.2 The *Contractor* keeps these records

- accounts of payments of Defined Cost,
- proof that the payments have been made,
- communications about and assessments of compensation events for Subcontractors and
- other records as stated in the Works Information.

52.3 The *Contractor* allows the *Project Manager* to inspect at any time within working hours the accounts and records which he is required to keep.

Providing the Works

20

20.2 The *Contractor* manages the *Contractor's* design, the provision of Site services and the construction and installation of the *works*. The *Contractor* subcontracts the *Contractor's* design, the provision of Site services and the construction and installation of the *works* except work which the Contract Data states that it will do.

20.3 The *Contractor* advises the *Project Manager* on the practical implications of the design of the *works* and on subcontracting arrangements.

20.4 The *Contractor* prepares forecasts of the total Defined Cost for the whole of the *works* in consultation with the *Project Manager* and submits them to the *Project Manager*. Forecasts are prepared at the intervals stated in the Contract Data from the *starting date* until Completion of the whole of the *works*. An explanation of the changes made since the previous forecast is submitted with each forecast.

Subcontracting

26

26.4 The *Contractor* submits the pricing information in the proposed subcontract documents for each subcontract to the *Project Manager* unless the *Project Manager* has agreed that no submission is required.

Assessing the amount due

50

50.8 Payments of Defined Cost made by the *Contractor* in a currency other than the *currency of the contract* are included in the amount due as payments to be made to it in the same currency. Such payments are converted to the *currency of the contract* in order to calculate the Fee using the *exchange rates*.

50.9 The *Contractor* notifies the *Project Manager* when a part of Defined Cost has been finalised, and makes available for inspection the records necessary to demonstrate that it has been correctly assessed. The *Project Manager* reviews the records made available, and no later than thirteen weeks after the *Contractor's* notification

- accepts that part of Defined Cost as correct,
- notifies the *Contractor* that further records are needed or
- notifies the *Contractor* of errors in its assessment.

The *Contractor* provides any further records requested or advises the correction of the errors in its assessment within four weeks of the *Project Manager's* notification. The *Project Manager* reviews the records provided, and within four weeks

- accepts the cost as correct or
- notifies the *Contractor* of the correct assessment of that part of Defined Cost.

If the *Project Manager* does not notify a decision on that part of Defined Cost within the time stated, the *Contractor's* assessment is treated as correct.

Defined Cost

52

52.3 The *Contractor* keeps these records

- accounts of payments made to Subcontractors,
- proof that the payments have been made,
- communications about and assessments of compensation events for Subcontractors and
- other records as stated in the Scope.

52.4 The *Contractor* allows the *Project Manager* to inspect at any time within working hours the accounts and records which it is required to keep.

Implementing compensation events

65

65.3 The changes to the forecast amount of the Prices, the Completion Date and the Key Dates are included in the notification implementing a compensation event.

20.5 If work which the *Contractor* is to do himself is affected by a compensation event, the *Project Manager* and the *Contractor* agree the change to the price for the work and any change to the Completion Date and Key Dates. If they cannot agree, the *Project Manager* decides the change.

Assessing compensation events

63

63.17 If work which the *Contractor* is to do is affected by a compensation event, the *Project Manager* and the *Contractor* agree the change to the price for the work and any change to the Completion Date and Key Dates. If they do not agree, the *Project Manager* decides the change.

GLOSSARY OF NEW ECC4 TERMS	
Client	changed project role – formerly '*Employer*'
Scope	changed defined term – formerly 'Works Information'
Dispute Avoidance Board	new entity, part of the revised dispute avoidance and resolution provisions (Option W3) (Option C)
People Rates	new defined term, part of new method for calculating the costs of people (Options A and B) – defined in clause 11.2(28)
people rates	new identified term
proposed subcontract documents	changed contract term (Options C, D, E, F) – formerly 'proposed contract data'
Short Schedule of Cost Components	changed title – formerly 'Shorter Schedule of Cost Components'
value engineering percentage	new identified term, used in assessing clause 16 *Contractor's* proposals (Options A and B)

DISPUTE RESOLUTION

Option W1

Dispute resolution procedure (used unless the United Kingdom Housing Grants, Construction and Regeneration Act 1996 applies).

Dispute resolution — **W1**

W1.1 A dispute arising under or in connection with this contract is referred to and decided by the *Adjudicator*.

Resolving and Avoiding Disputes

Note that the title of this section of the contract has changed. The drafters are stressing that the Parties should try to avoid disputes wherever possible. 

OPTION W1

Used when Adjudication is the method of dispute resolution and the United Kingdom Housing Grants, Construction and Regeneration Act 1996 does not apply.

Resolving disputes **W1**

W1.1 (1) A dispute arising under or in connection with the contract is referred to the *Senior Representatives* in accordance with the Dispute Reference Table. If the dispute is not resolved by the *Senior Representatives*, it is referred to and decided by the *Adjudicator*.

(2) The Party referring a dispute notifies the *Senior Representatives*, the other Party and the *Project Manager* of the nature of the dispute it wishes to resolve. Each Party submits to the other their statement of case within one week of the notification. Each statement of case is limited to no more than ten sheets of A4 paper together with supporting evidence, unless otherwise agreed by the Parties.

(3) The *Senior Representatives* attend as many meetings and use any procedure they consider necessary to try to resolve the dispute over a period of no more than three weeks. At the end of this period the *Senior Representatives* produce a list of the issues agreed and issues not agreed. The *Project Manager* and the *Contractor* put into effect the issues agreed.

(4) No evidence of the statement of case or discussions is disclosed, used or referred to in any subsequent proceedings before the *Adjudicator* or the *tribunal*.

DISPUTE REFERENCE TABLE		
DISPUTE ABOUT	**WHICH PARTY MAY REFER IT TO THE *SENIOR REPRESENTATIVES*?**	**WHEN MAY IT BE REFERRED TO THE *SENIOR REPRESENTATIVES*?**
An action or inaction of the *Project Manager* or the *Supervisor*	Either Party	Not more than four weeks after the Party becomes aware of the action or inaction
A programme, compensation event or quotation for a compensation event which is treated as having been accepted	The *Client*	Not more than four weeks after it was treated as accepted
An assessment of Defined Cost which is treated as correct	Either Party	Not more than four weeks after the assessment was treated as correct
Any other matter	Either Party	When the dispute arises

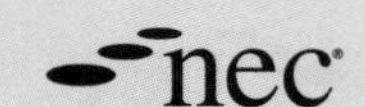

The *Adjudicator* W1.2

(1) The Parties appoint the *Adjudicator* under the NEC Adjudicator's Contract current at the *starting date*.

(2) The *Adjudicator* acts impartially and decides the dispute as an independent adjudicator and not as an arbitrator.

(3) If the *Adjudicator* is not identified in the Contract Data or if the *Adjudicator* resigns or is unable to act, the Parties choose a new adjudicator jointly. If the Parties have not chosen an adjudicator, either Party may ask the *Adjudicator nominating body* to choose one. The *Adjudicator nominating body* chooses an adjudicator within four days of the request. The chosen adjudicator becomes the *Adjudicator*.

(4) A replacement *Adjudicator* has the power to decide a dispute referred to his predecessor but not decided at the time when the predecessor resigned or became unable to act. He deals with an undecided dispute as if it had been referred to him on the date he was appointed.

(5) The *Adjudicator*, his employees and agents are not liable to the Parties for any action or failure to take action in an adjudication unless the action or failure to take action was in bad faith.

The adjudication W1.3

(1) Disputes are notified and referred to the *Adjudicator* in accordance with the Adjudication Table.

ADJUDICATION TABLE

Dispute about	**Which Party may refer it to the *Adjudicator*?**	**When may it be referred to the *Adjudicator*?**
An action of the *Project Manager* or the *Supervisor*	The *Contractor*	Between two and four weeks after the *Contractor*'s notification of the dispute to the *Employer* and the *Project Manager*, the notification itself being made not more than four weeks after the *Contractor* becomes aware of the action
The *Project Manager* or *Supervisor* not having taken an action	The *Contractor*	Between two and four weeks after the *Contractor*'s notification of the dispute to the *Employer* and the *Project Manager*, the notification itself being made not more than four weeks after the *Contractor* becomes aware that the action was not taken
A quotation for a compensation event which is treated as having been accepted	The *Employer*	Between two and four weeks after the *Project Manager*'s notification of the dispute to the *Employer* and the *Contractor*, the notification itself being made not more than four weeks after the quotation was treated as accepted
Any other matter	Either Party	Between two and four weeks after notification of the dispute to the other Party and the *Project Manager*

The main change to Option W1 is the introduction of a new dispute resolution procedure involving *Senior Representatives* from the Parties. This is a mandatory requirement of Option W1. The *Senior Representatives* in turn can use whatever procedures they deem necessary within the relatively short time period allowed to try to resolve the dispute in a timely and cost-effective manner. The other procedures might include negotiation, mediation or the like. Any issues not agreed can then be taken to adjudication, if either Party still wishes to do so. Sometimes, a final sanity check such as this might just result in the issues being agreed before the dispute formalises and takes a very different direction. The Dispute Reference Table has been modified also, allowing either Party to refer any dispute to the *Senior Representatives,* with the exception of the second provision, which is only available to the *Client*.

The *Adjudicator* W1.2

(1) The Parties appoint the *Adjudicator* under the NEC Dispute Resolution Service Contract current at the *starting date*.

(2) The *Adjudicator* acts impartially and decides the dispute as an independent adjudicator and not as an arbitrator.

(3) If the *Adjudicator* is not identified in the Contract Data or if the *Adjudicator* resigns or is unable to act, the Parties choose a new adjudicator jointly. If the Parties have not chosen an adjudicator, either Party may ask the *Adjudicator nominating body* to choose one. The *Adjudicator nominating body* chooses an adjudicator within seven days of the request. The chosen adjudicator becomes the *Adjudicator*.

(4) A replacement *Adjudicator* has the power to decide a dispute referred to a predecessor but not decided at the time when the predecessor resigned or became unable to act. The *Adjudicator* deals with an undecided dispute as if it had been referred on the date of appointment as replacement *Adjudicator*.

(5) The *Adjudicator* and the *Adjudicator's* employees and agents are not liable to the Parties for any action or failure to take action in an adjudication unless the action or failure to take action was in bad faith.

The adjudication W1.3

(1) A Party disputing any issue not agreed by the *Senior Representatives* issues a notice of adjudication to the other Party and the *Project Manager* within two weeks of the production of the list of agreed and not agreed issues, or when it should have been produced. The dispute is referred to the *Adjudicator* within one week of the notice of adjudication.

(2) The times for notifying and referring a dispute may be extended by the *Project Manager* if the *Contractor* and the *Project Manager* agree to the extension before the notice or referral is due. The *Project Manager* informs the *Contractor* of the extension that has been agreed. If a disputed matter is not notified and referred within the times set out in the contract, neither Party may subsequently refer it to the *Adjudicator* or the *tribunal*.

(3) The Party referring the dispute to the *Adjudicator* includes with its referral information to be considered by the *Adjudicator*. Any more information from a Party to be considered by the *Adjudicator* is provided within four weeks of the referral. This period may be extended if the *Adjudicator* and the Parties agree.

(4) If a matter disputed by the *Contractor* under or in connection with a subcontract is also a matter disputed under or in connection with the contract and if the subcontract allows, the *Contractor* may refer the subcontract dispute to the *Adjudicator* at the same time as the main contract referral. The *Adjudicator* then decides the disputes together and references to the Parties for the purposes of the dispute are interpreted as including the Subcontractor.

(2) The times for notifying and referring a dispute may be extended by the *Project Manager* if the *Contractor* and the *Project Manager* agree to the extension before the notice or referral is due. The *Project Manager* notifies the extension that has been agreed to the *Contractor.* If a disputed matter is not notified and referred within the times set out in this contract, neither Party may subsequently refer it to the *Adjudicator* or the *tribunal*.

(3) The Party referring the dispute to the *Adjudicator* includes with his referral information to be considered by the *Adjudicator*. Any more information from a Party to be considered by the *Adjudicator* is provided within four weeks of the referral. This period may be extended if the *Adjudicator* and the Parties agree.

(4) If a matter disputed by the *Contractor* under or in connection with a subcontract is also a matter disputed under or in connection with this contract and if the subcontract allows, the *Contractor* may refer the subcontract dispute to the *Adjudicator* at the same time as the main contract referral. The *Adjudicator* then decides the disputes together and references to the Parties for the purposes of the dispute are interpreted as including the Subcontractor.

(5) The *Adjudicator* may

- review and revise any action or inaction of the *Project Manager* or *Supervisor* related to the dispute and alter a quotation which has been treated as having been accepted,
- take the initiative in ascertaining the facts and the law related to the dispute,
- instruct a Party to provide further information related to the dispute within a stated time and
- instruct a Party to take any other action which he considers necessary to reach his decision and to do so within a stated time.

(6) A communication between a Party and the *Adjudicator* is communicated to the other Party at the same time.

(7) If the *Adjudicator's* decision includes assessment of additional cost or delay caused to the *Contractor*, he makes his assessment in the same way as a compensation event is assessed.

(8) The *Adjudicator* decides the dispute and notifies the Parties and the *Project Manager* of his decision and his reasons within four weeks of the end of the period for receiving information. This four week period may be extended if the Parties agree.

(9) Unless and until the *Adjudicator* has notified the Parties of his decision, the Parties, the *Project Manager* and the *Supervisor* proceed as if the matter disputed was not disputed.

(10) The *Adjudicator's* decision is binding on the Parties unless and until revised by the *tribunal* and is enforceable as a matter of contractual obligation between the Parties and not as an arbitral award. The *Adjudicator's* decision is final and binding if neither Party has notified the other within the times required by this contract that he is dissatisfied with a decision of the *Adjudicator* and intends to refer the matter to the *tribunal*.

(11) The *Adjudicator* may, within two weeks of giving his decision to the Parties, correct any clerical mistake or ambiguity.

(5) The *Adjudicator* may

- review and revise any action or inaction of the *Project Manager* or *Supervisor* related to the dispute and alter a matter which has been treated as accepted or correct,
- take the initiative in ascertaining the facts and the law related to the dispute,
- instruct a Party to provide further information related to the dispute within a stated time and
- instruct a Party to take any other action which is considered necessary for the *Adjudicator* to reach a decision and to do so within a stated time.

(6) A communication between a Party and the *Adjudicator* is communicated to the other Party at the same time.

(7) If the *Adjudicator's* decision includes assessment of additional cost or delay caused to the *Contractor*, the assessment is made in the same way as a compensation event is assessed.

(8) The *Adjudicator* decides the dispute and informs the Parties and the *Project Manager* of the decision and reasons within four weeks of the end of the period for receiving information. This four week period may be extended if the Parties agree.

(9) Unless and until the *Adjudicator* has informed the Parties of the decision, the Parties, the *Project Manager* and the *Supervisor* proceed as if the matter disputed was not disputed.

(10) The *Adjudicator's* decision is binding on the Parties unless and until revised by the *tribunal* and is enforceable as a matter of contractual obligation between the Parties and not as an arbitral award. The *Adjudicator's* decision is final and binding if neither Party has notified the other within the times required by the contract that it is dissatisfied with a decision of the *Adjudicator* and intends to refer the matter to the *tribunal*. A Party does not refer a dispute to the *Adjudicator* that is the same or substantially the same as one that has already been referred to the *Adjudicator*.

The Parties cannot keep going back to adjudication over the same issue in an attempt to achieve a different outcome.

(11) The *Adjudicator* may, within two weeks of giving the decision to the Parties, correct any clerical mistake or ambiguity.

Review by the *tribunal*

W1.4 (1) A Party does not refer any dispute under or in connection with this contract to the *tribunal* unless it has first been referred to the *Adjudicator* in accordance with this contract.

(2) If, after the *Adjudicator* notifies his decision a Party is dissatisfied, he may notify the other Party that he intends to refer it to the *tribunal*. A Party may not refer a dispute to the *tribunal* unless this notification is given within four weeks of notification of the *Adjudicator*'s decision.

(3) If the *Adjudicator* does not notify his decision within the time provided by this contract, a Party may notify the other Party that he intends to refer the dispute to the *tribunal*. A Party may not refer a dispute to the *tribunal* unless this notification is given within four weeks of the date by which the *Adjudicator* should have notified his decision.

(4) The *tribunal* settles the dispute referred to it. The *tribunal* has the powers to reconsider any decision of the *Adjudicator* and review and revise any action or inaction of the *Project Manager* or the *Supervisor* related to the dispute. A Party is not limited in the *tribunal* proceedings to the information, evidence or arguments put to the *Adjudicator*.

(5) If the *tribunal* is arbitration, the *arbitration procedure*, the place where the arbitration is to be held and the method of choosing the arbitrator are those stated in the Contract Data.

(6) A Party does not call the *Adjudicator* as a witness in *tribunal* proceedings.

The *tribunal* W1.4

(1) A Party does not refer any dispute under or in connection with the contract to the *tribunal* unless it has first been referred to the *Adjudicator* in accordance with the contract.

(2) If, after being informed of the *Adjudicator's* decision, a Party is dissatisfied, that Party may notify the other Party of the matter which is disputed and state that it intends to refer the matter to the *tribunal*. The dispute is not referred to the *tribunal* unless this notification is given within four weeks of being informed of the *Adjudicator's* decision.

(3) If the *Adjudicator* does not inform the Parties of the decision within the time provided by the contract, a Party may notify the other Party that it intends to refer the dispute to the *tribunal*. A Party does not refer a dispute to the *tribunal* unless this notification is given within four weeks of the date by which the *Adjudicator* should have informed the Parties of the decision.

(4) The *tribunal* settles the dispute referred to it. The *tribunal* has the powers to reconsider any decision of the *Adjudicator* and review and revise any action or inaction of the *Project Manager* or the *Supervisor* related to the dispute. A Party is not limited in the *tribunal* proceedings to the information, evidence or arguments put to the *Adjudicator*.

(5) If the *tribunal* is arbitration, the *arbitration procedure*, the place where the arbitration is to be held and the method of choosing the arbitrator are those stated in the Contract Data.

(6) A Party does not call the *Adjudicator* as a witness in *tribunal* proceedings.

Option W2

Dispute resolution procedure (used in the United Kingdom when the Housing Grants, Construction and Regeneration Act 1996 applies).

OPTION W2

Used when Adjudication is the method of dispute resolution and the United Kingdom Housing Grants, Construction and Regeneration Act 1996 applies.

Resolving disputes **W2**

W2.1 (1) If the Parties agree, a dispute arising under or in connection with the contract is referred to the *Senior Representatives*. If the dispute is not resolved by the *Senior Representatives*, it is referred to and decided by the *Adjudicator*.

(2) The Party referring a dispute notifies the *Senior Representatives*, the other Party and the *Project Manager* of the nature of the dispute it wishes to resolve. Each Party submits to the other their statement of case within one week of the notification. Each statement of case is limited to no more than ten sheets of A4 paper together with supporting evidence, unless otherwise agreed by the Parties.

(3) The *Senior Representatives* attend as many meetings and use any procedure they consider necessary to try to resolve the dispute over a period of up to three weeks. At the end of this period the *Senior Representatives* produce a list of the issues agreed and issues not agreed. The *Project Manager* and the *Contractor* put into effect the issues agreed.

(4) No evidence of the statement of case or discussions is disclosed, used or referred to in any subsequent proceedings before the *Adjudicator* or the *tribunal*.

Option W2 also incorporates the new dispute resolution procedure, where disputes are referred first to the *Senior Representatives*. In this Option though, and unlike Option W1, the procedure is written to comply with the Housing Grants, Construction and Regeneration Act 1996, in that the *Senior Representatives* only get involved if the Parties agree. The rights of both Parties to be able to go to adjudication are preserved.

Dispute resolution

W2

W2.1 (1) A dispute arising under or in connection with this contract is referred to and decided by the *Adjudicator*. A Party may refer a dispute to the *Adjudicator* at any time.

(2) In this Option, time periods stated in days exclude Christmas Day, Good Friday and bank holidays.

The *Adjudicator*

W2.2 (1) The Parties appoint the *Adjudicator* under the NEC Adjudicator's Contract current at the *starting date*.

(2) The *Adjudicator* acts impartially and decides the dispute as an independent adjudicator and not as an arbitrator.

(3) If the *Adjudicator* is not identified in the Contract Data or if the *Adjudicator* resigns or becomes unable to act

- the Parties may choose an adjudicator jointly or
- a Party may ask the *Adjudicator nominating body* to choose an adjudicator.

The *Adjudicator nominating body* chooses an adjudicator within four days of the request. The chosen adjudicator becomes the *Adjudicator.*

(4) A replacement *Adjudicator* has the power to decide a dispute referred to his predecessor but not decided at the time when his predecessor resigned or became unable to act. He deals with an undecided dispute as if it had been referred to him on the date he was appointed.

(5) The *Adjudicator,* his employees and agents are not liable to the Parties for any action or failure to take action in an adjudication unless the action or failure to take action was in bad faith.

The *Adjudicator* W2.2

(1) A dispute arising under or in connection with the contract is referred to and decided by the *Adjudicator*. A Party may refer a dispute to the *Adjudicator* at any time whether or not the dispute has been referred to the *Senior Representatives*.

(2) In this Option, time periods stated in days exclude Christmas Day, Good Friday and bank holidays.

(3) The Parties appoint the *Adjudicator* under the NEC Dispute Resolution Service Contract current at the *starting date*.

(4) The *Adjudicator* acts impartially and decides the dispute as an independent adjudicator and not as an arbitrator.

(5) If the *Adjudicator* is not identified in the Contract Data or if the *Adjudicator* resigns or becomes unable to act

- the Parties may choose an adjudicator jointly or
- a Party may ask the *Adjudicator nominating body* to choose an adjudicator.

The *Adjudicator nominating body* chooses an adjudicator within four days of the request. The chosen adjudicator becomes the *Adjudicator*.

(6) A replacement *Adjudicator* has the power to decide a dispute referred to a predecessor but not decided at the time when the predecessor resigned or became unable to act. The *Adjudicator* deals with an undecided dispute as if it had been referred on the date of appointment as replacement *Adjudicator*.

(7) A Party does not refer a dispute to the *Adjudicator* that is the same or substantially the same as one that has already been decided by the *Adjudicator*.

(8) The *Adjudicator*, and the *Adjudicator's* employees and agents are not liable to the Parties for any action or failure to take action in an adjudication unless the action or failure to take action was in bad faith.

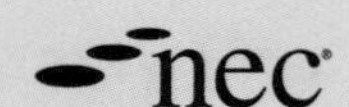

The adjudication

W2.3 (1) Before a Party refers a dispute to the *Adjudicator*, he gives a notice of adjudication to the other Party with a brief description of the dispute and the decision which he wishes the *Adjudicator* to make. If the *Adjudicator* is named in the Contract Data, the Party sends a copy of the notice of adjudication to the *Adjudicator* when it is issued. Within three days of the receipt of the notice of adjudication, the *Adjudicator* notifies the Parties

- that he is able to decide the dispute in accordance with the contract or
- that he is unable to decide the dispute and has resigned.

If the *Adjudicator* does not so notify within three days of the issue of the notice of adjudication, either Party may act as if he has resigned.

(2) Within seven days of a Party giving a notice of adjudication he

- refers the dispute to the *Adjudicator*,
- provides the *Adjudicator* with the information on which he relies, including any supporting documents and
- provides a copy of the information and supporting documents he has provided to the *Adjudicator* to the other Party.

Any further information from a Party to be considered by the *Adjudicator* is provided within fourteen days of the referral. This period may be extended if the *Adjudicator* and the Parties agree.

(3) If a matter disputed by the *Contractor* under or in connection with a subcontract is also a matter disputed under or in connection with this contract, the *Contractor* may, with the consent of the Subcontractor, refer the subcontract dispute to the *Adjudicator* at the same time as the main contract referral. The *Adjudicator* then decides the disputes together and references to the Parties for the purposes of the dispute are interpreted as including the Subcontractor.

(4) The *Adjudicator* may

- review and revise any action or inaction of the *Project Manager* or *Supervisor* related to the dispute and alter a quotation which has been treated as having been accepted,
- take the initiative in ascertaining the facts and the law related to the dispute,
- instruct a Party to provide further information related to the dispute within a stated time and
- instruct a Party to take any other action which he considers necessary to reach his decision and to do so within a stated time.

(5) If a Party does not comply with any instruction within the time stated by the *Adjudicator*, the *Adjudicator* may continue the adjudication and make his decision based upon the information and evidence he has received.

(6) A communication between a Party and the *Adjudicator* is communicated to the other Party at the same time.

The adjudication W2.3

(1) Before a Party refers a dispute to the *Adjudicator*, it gives a notice of adjudication to the other Party with a brief description of the dispute and the decision which it wishes the *Adjudicator* to make. If the *Adjudicator* is named in the Contract Data, the Party sends a copy of the notice of adjudication to the *Adjudicator* when it is issued. Within three days of the receipt of the notice of adjudication, the *Adjudicator* informs the Parties that the *Adjudicator*

- is able to decide the dispute in accordance with the contract or
- is unable to decide the dispute and has resigned.

If the *Adjudicator* does not so inform within three days of the issue of the notice of adjudication, either Party may act as if the *Adjudicator* has resigned.

(2) Within seven days of a Party giving a notice of adjudication it

- refers the dispute to the *Adjudicator*,
- provides the *Adjudicator* with the information on which it relies, including any supporting documents and
- provides a copy of the information and supporting documents it has provided to the *Adjudicator* to the other Party.

Any further information from a Party to be considered by the *Adjudicator* is provided within fourteen days of the referral. This period may be extended if the *Adjudicator* and the Parties agree.

(3) If a matter disputed by the *Contractor* under or in connection with a subcontract is also a matter disputed under or in connection with the contract, the *Contractor* may, with the consent of the Subcontractor, refer the subcontract dispute to the *Adjudicator* at the same time as the main contract referral. The *Adjudicator* then decides the disputes together and references to the Parties for the purposes of the dispute are interpreted as including the Subcontractor.

(4) The *Adjudicator* may

- review and revise any action or inaction of the *Project Manager* or *Supervisor* related to the dispute and alter a matter which has been treated as accepted or correct,
- take the initiative in ascertaining the facts and the law related to the dispute,
- instruct a Party to provide further information related to the dispute within a stated time and
- instruct a Party to take any other action which is considered necessary to reach a decision and to do so within a stated time.

(5) If a Party does not comply with any instruction within the time stated by the *Adjudicator*, the *Adjudicator* may continue the adjudication and make a decision based upon the information and evidence received.

(6) A communication between a Party and the *Adjudicator* is communicated to the other Party at the same time.

(7) If the *Adjudicator*'s decision includes assessment of additional cost or delay caused to the *Contractor*, he makes his assessment in the same way as a compensation event is assessed. If the *Adjudicator*'s decision changes an amount notified as due, payment of the sum decided by the *Adjudicator* is due not later than seven days from the date of the decision or the final date for payment of the notified amount, whichever is the later.

(8) The *Adjudicator* decides the dispute and notifies the Parties and the *Project Manager* of his decision and his reasons within twenty-eight days of the dispute being referred to him. This period may be extended by up to fourteen days with the consent of the referring Party or by any other period agreed by the Parties. The *Adjudicator* may in his decision allocate his fees and expenses between the Parties.

(9) Unless and until the *Adjudicator* has notified the Parties of his decision, the Parties, the *Project Manager* and the *Supervisor* proceed as if the matter disputed was not disputed.

(10) If the *Adjudicator* does not make his decision and notify it to the Parties within the time provided by this contract, the Parties and the *Adjudicator* may agree to extend the period for making his decision. If they do not agree to an extension, either Party may act as if the *Adjudicator* has resigned.

(11) The *Adjudicator*'s decision is binding on the Parties unless and until revised by the *tribunal* and is enforceable as a matter of contractual obligation between the Parties and not as an arbitral award. The *Adjudicator*'s decision is final and binding if neither Party has notified the other within the times required by this contract that he is dissatisfied with a matter decided by the *Adjudicator* and intends to refer the matter to the *tribunal*.

(12) The *Adjudicator* may, within five days of giving his decision to the Parties, correct the decision to remove a clerical or typographical error arising by accident or omission.

Review by the *tribunal* W2.4

(1) A Party does not refer any dispute under or in connection with this contract to the *tribunal* unless it has first been decided by the *Adjudicator* in accordance with this contract.

(2) If, after the *Adjudicator* notifies his decision a Party is dissatisfied, that Party may notify the other Party of the matter which he disputes and state that he intends to refer it to the *tribunal*. The dispute may not be referred to the *tribunal* unless this notification is given within four weeks of the notification of the *Adjudicator*'s decision.

(3) The *tribunal* settles the dispute referred to it. The *tribunal* has the powers to reconsider any decision of the *Adjudicator* and to review and revise any action or inaction of the *Project Manager* or the *Supervisor* related to the dispute. A Party is not limited in *tribunal* proceedings to the information or evidence put to the *Adjudicator*.

(4) If the *tribunal* is arbitration, the *arbitration procedure,* the place where the arbitration is to be held and the method of choosing the arbitrator are those stated in the Contract Data.

(5) A Party does not call the *Adjudicator* as a witness in *tribunal* proceedings.

(7) If the *Adjudicator's* decision includes assessment of additional cost or delay caused to the *Contractor,* the assessment is made in the same way as a compensation event is assessed. If the *Adjudicator's* decision changes an amount notified as due, the date on which payment of the changed amount becomes due is seven days after the date of the decision.

(8) The *Adjudicator* decides the dispute and informs the Parties and the *Project Manager* of the decision and reasons within twenty eight days of the dispute being referred. This period may be extended by up to fourteen days with the consent of the referring Party or by any other period agreed by the Parties. The *Adjudicator* may in the decision allocate the *Adjudicator's* fees and expenses between the Parties.

(9) Unless and until the *Adjudicator* has informed the Parties of the decision, the Parties, the *Project Manager* and the *Supervisor* proceed as if the matter disputed was not disputed.

(10) If the *Adjudicator* does not inform the Parties of the decision within the time provided by the contract, the Parties and the *Adjudicator* may agree to extend the period for making a decision. If they do not agree to an extension, either Party may act as if the *Adjudicator* has resigned.

(11) The *Adjudicator's* decision is binding on the Parties unless and until revised by the *tribunal* and is enforceable as a matter of contractual obligation between the Parties and not as an arbitral award. The *Adjudicator's* decision is final and binding if neither Party has notified the other within the times required by the contract that it is dissatisfied with a matter decided by the *Adjudicator* and intends to refer the matter to the *tribunal.*

(12) The *Adjudicator* may, within five days of giving the decision to the Parties, correct the decision to remove a clerical or typographical error arising by accident or omission.

The *tribunal* W2.4

(1) A Party does not refer any dispute under or in connection with the contract to the *tribunal* unless it has first been decided by the *Adjudicator* in accordance with the contract.

(2) If, after the *Adjudicator* makes a decision a Party is dissatisfied, that Party may notify the other Party of the matter which is disputed and state that it intends to refer the disputed matter to the *tribunal.* The dispute may not be referred to the *tribunal* unless this notification is given within four weeks of being informed of the *Adjudicator's* decision.

(3) The *tribunal* settles the dispute referred to it. The *tribunal* has the power to reconsider any decision of the *Adjudicator* and to review and revise any action or inaction of the *Project Manager* or the *Supervisor* related to the dispute. A Party is not limited in *tribunal* proceedings to the information, or evidence or arguments put to the *Adjudicator.*

(4) If the *tribunal* is arbitration, the *arbitration procedure,* the place where the arbitration is to be held and the method of choosing the arbitrator are those stated in the Contract Data.

(5) A Party does not call the *Adjudicator* as a witness in *tribunal* proceedings.

OPTION W3

Procedure for resolving and avoiding disputes (either W1 or W3 is used unless the United Kingdom Housing Grants, Construction and Regeneration Act 1996 applies in which case W2 applies).

The Dispute Avoidance Board W3

W3.1 (1) The Dispute Avoidance Board consists of one or three members as identified in the Contract Data. If the Contract Data states that the number of members is three, the third member is jointly chosen by the Parties.

(2) The Parties appoint the Dispute Avoidance Board under the NEC Dispute Resolution Service Contract current at the *starting date*.

(3) The Dispute Avoidance Board act impartially.

(4) If a member of the Dispute Avoidance Board is not identified in the Contract Data or if a member of the Dispute Avoidance Board is unable to act, the Parties jointly choose a new member. If the Parties have not chosen a Dispute Avoidance Board member or a replacement, either Party may ask the *Dispute Avoidance Board nominating body* to choose one. The *Dispute Avoidance Board nominating body* choose a Dispute Avoidance Board member within seven days of the request. The chosen member becomes a member of the Dispute Avoidance Board.

(5) The Dispute Avoidance Board visit the Site at the intervals stated in the Contract Data from the *starting date* until the *defects date* unless the Parties agree that a visit is not necessary. The purpose of the visit is to enable the Dispute Avoidance Board to inspect the progress of the *works* and become aware of any potential disputes. The Dispute Avoidance Board make additional visits when requested by the Parties.

(6) The agenda for the site visit is proposed by the Parties and decided by the Dispute Avoidance Board.

(7) The members of the Dispute Avoidance Board, their employees and agents are not liable to the Parties for any action or failure to take action in resolving a potential dispute unless the action or failure to take action was in bad faith.

Resolving potential disputes W3.2

(1) The Dispute Avoidance Board assists the Parties in resolving potential disputes before they become disputes.

(2) A potential dispute arising under or in connection with the contract is referred to the Dispute Avoidance Board.

(3) Potential disputes are notified and referred to the Dispute Avoidance Board between two to four weeks after notification of the issue to the other Party and the *Project Manager*.

(4) The Parties make available to the Dispute Avoidance Board

- copies of the contract,
- progress reports and
- any other material they consider relevant to any difference which they wish the Dispute Avoidance Board to consider in advance of the visit to the Site.

(5) The Dispute Avoidance Board

- visit the Site and inspect the works,
- review all potential disputes and help the Parties to settle them without the need for the dispute to be formally referred,
- prepare a note of their visit and
- unless the Parties have resolved the potential dispute by the end of the Site visit, provide a recommendation for resolving it.

(6) The Dispute Avoidance Board can take the initiative in reviewing potential disputes, including asking the Parties to provide further information.

The *tribunal* W3.3

(1) A Party does not refer any dispute under or in connection with the contract to the *tribunal* unless it has first been referred to the Dispute Avoidance Board as a potential dispute in accordance with the contract.

(2) If, after the Dispute Avoidance Board makes a recommendation a Party is dissatisfied, that Party may notify the other Party of the matter which it disputes and state that it intends to refer it to the *tribunal*. The dispute is not referred to the *tribunal* unless this notification is given within four weeks of notification of the Dispute Avoidance Board's recommendation.

(3) The *tribunal* settles the dispute referred to it. The *tribunal* has the powers to reconsider any recommendation of the Dispute Avoidance Board and review and revise any action or inaction of the *Project Manager* or the *Supervisor* related to the dispute. A Party is not limited in the *tribunal* proceedings to the information, evidence or arguments put to the Dispute Avoidance Board.

(4) If the *tribunal* is arbitration, the *arbitration* procedure, the place where the arbitration is to be held and the method of choosing the arbitrator are those stated in the Contract Data.

(5) A Party does not call a member of the Dispute Avoidance Board as a witness in *tribunal* proceedings.

Option W3 is a new provision to some of the NEC4 contracts. It is most likely to be used on very substantial projects or programmes of work. Note that this is a Dispute Avoidance Board and not a Dispute Resolution Board. Again, the drafters have put the emphasis on dispute avoidance.

The Dispute Avoidance Board (DAB) comprises either one or three members; if three then the final member is jointly chosen by the Parties after the Contract Date. The DAB is appointed using the NEC Dispute Resolution Service Contract (a re-vamped version of the previous NEC Adjudicator's Contract), which has been drafted to reflect the use of other dispute resolution services, not just adjudication. The DAB visits the Site regularly, unless the Parties decide otherwise, and its main task is to assist the Parties in resolving potential disputes before they become disputes. The DAB can take the initiative here, and the Parties cannot go to the *tribunal* until after they have taken the matter to the DAB as a potential dispute.

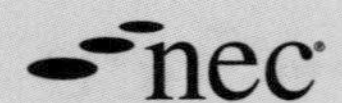

SECONDARY OPTION CLAUSES

Option X1: Price adjustment for inflation (used only with Options A, B, C and D)

Defined terms	**X1**	
	X1.1	(a) The Base Date Index (B) is the latest available index before the *base date*. (b) The Latest Index (L) is the latest available index before the date of assessment of an amount due. (c) The Price Adjustment Factor is the total of the products of each of the proportions stated in the Contract Data multiplied by (L – B)/B for the index linked to it.
Price Adjustment Factor	X1.2	If an index is changed after it has been used in calculating a Price Adjustment Factor, the calculation is repeated and a correction included in the next assessment of the amount due. The Price Adjustment Factor calculated at the Completion Date for the whole of the *works* is used for calculating price adjustment after this date.
Price adjustment Options A and B	X1.4	Each amount due includes an amount for price adjustment which is the sum of • the change in the Price for Work Done to Date since the last assessment of the amount due multiplied by the Price Adjustment Factor for the date of the current assessment, • the amount for price adjustment included in the previous amount due and • correcting amounts, not included elsewhere, which arise from changes to indices used for assessing previous amounts for price adjustment.
Price adjustment Options C and D	X1.5	Each time the amount due is assessed, an amount for price adjustment is added to the total of the Prices which is the sum of • the change in the Price for Work Done to Date since the last assessment of the amount due multiplied by (PAF/(1+PAF)) where PAF is the Price Adjustment Factor for the date of the current assessment and • correcting amounts, not included elsewhere, which arise from changes to indices used for assessing previous amounts for price adjustment.
Compensation events	X1.3	The Defined Cost for compensation events is assessed using the • Defined Cost current at the time of assessing the compensation event adjusted to base date by dividing by one plus the Price Adjustment Factor for the last assessment of the amount due and • Defined Cost at base date levels for amounts calculated from rates stated in the Contract Data for employees and Equipment.

Secondary Option Clauses

OPTION X1: PRICE ADJUSTMENT FOR INFLATION (USED ONLY WITH OPTIONS A, B, C AND D)

Defined terms	X1 X1.1	(a) The Base Date Index (B) is the latest available index before the *base date*. (b) The Latest Index (L) is the latest available index before the date of assessment of an amount due. (c) The Price Adjustment Factor (PAF) at each date of assessment of an amount due is the total of the products of each of the proportions stated in the Contract Data multiplied by (L – B)/B for the index linked to it.
Price Adjustment Factor	X1.2	If an index is changed after it has been used in calculating a PAF, the calculation is not changed. The PAF calculated at the last assessment date before the Completion Date for the whole of the *works* is used for calculating an amount for price adjustment after that date.
Price adjustment Options A and B	X1.3	Each amount due includes an amount for price adjustment which is the sum of • the change in the Price for Work Done to Date since the last assessment of the amount due multiplied by the PAF and • the amount for price adjustment included in the previous amount due.
Price adjustment Options C and D	X1.4	Each time the amount due is assessed, an amount for price adjustment is added to the total of the Prices which is the change in the Price for Work Done to Date since the last assessment of the amount due multiplied by (PAF/(1+PAF)).
Compensation events	X1.5	The Defined Cost for compensation events is assessed using • the Defined Cost at *base date* levels for amounts calculated from rates stated in the Contract Data for people and Equipment and • the Defined Cost current at the dividing date used in assessing the compensation event, adjusted to the *base date* by dividing by one plus the PAF for the last assessment of the amount due before that dividing date, for other amounts.

Option X2: Changes in the law

Changes in the law

X2

X2.1 A change in the law of the country in which the Site is located is a compensation event if it occurs after the Contract Date. The *Project Manager* may notify the *Contractor* of a compensation event for a change in the law and instruct him to submit quotations. If the effect of a compensation event which is a change in the law is to reduce the total Defined Cost, the Prices are reduced.

Option X3: Multiple currencies (used only with Options A and B)

Multiple currencies

X3

X3.1 The *Contractor* is paid in currencies other than the *currency of this contract* for the items or activities listed in the Contract Data. The *exchange rates* are used to convert from the *currency of this contract* to other currencies.

X3.2 Payments to the *Contractor* in currencies other than the *currency of this contract* do not exceed the maximum amounts stated in the Contract Data. Any excess is paid in the *currency of this contract*.

Option X4: Parent company guarantee

Parent company guarantee

X4

X4.1 If a parent company owns the *Contractor*, the *Contractor* gives to the *Employer* a guarantee by the parent company of the *Contractor*'s performance in the form set out in the Works Information. If the guarantee was not given by the Contract Date, it is given to the *Employer* within four weeks of the Contract Date.

OPTION X2: CHANGES IN THE LAW

Changes in the law X2

X2.1 A change in the law of the country in which the Site is located is a compensation event if it occurs after the Contract Date. If the effect of a compensation event which is a change in the law is to reduce the total Defined Cost, the Prices are reduced.

OPTION X3: MULTIPLE CURRENCIES (USED ONLY WITH OPTIONS A AND B)

Multiple currencies X3

X3.1 The *Contractor* is paid in currencies other than the *currency of the contract* for the items or activities listed in the Contract Data. The *exchange rates* are used to convert from the *currency of the contract* to other currencies.

X3.2 Payments to the *Contractor* in currencies other than the *currency of the contract* do not exceed the maximum amounts stated in the Contract Data. Any excess is paid in the *currency of the contract*.

OPTION X4: ULTIMATE HOLDING COMPANY GUARANTEE

Ultimate holding company guarantee X4

X4.1 If the *Contractor* is a subsidiary of another company, the *Contractor* gives to the *Client* a guarantee of the *Contractor's* performance from the ultimate holding company of the *Contractor* in the form set out in the Scope. If the guarantee was not given by the Contract Date, it is given to the *Client* within four weeks of the Contract Date.

X4.2 The *Contractor* may propose an alternative guarantor who is also owned by the ultimate holding company for acceptance by the *Project Manager*. A reason for not accepting the guarantor is that their commercial position is not strong enough to carry the guarantee.

The wording in Option X4 is changed so that the guarantee is secured from the ultimate holding company, rather than from the parent company. The new provision in clause X4.2 provides for the *Contractor* to propose an alternative to this. 

Option X5: Sectional Completion

Sectional Completion

X5

X5.1 In these *conditions of contract*, unless stated as the whole of the *works*, each reference and clause relevant to

- the *works*,
- Completion and
- Completion Date

applies, as the case may be, to either the whole of the *works* or any *section* of the *works*.

Option X6: Bonus for early Completion

Bonus for early Completion

X6

X6.1 The *Contractor* is paid a bonus calculated at the rate stated in the Contract Data for each day from the earlier of

- Completion and
- the date on which the *Employer* takes over the *works*

until the Completion Date.

OPTION X5: SECTIONAL COMPLETION

Sectional Completion

X5

X5.1 In these *conditions of contract*, unless stated as the whole of the *works*, each reference and clause relevant to

- the *works*,
- Completion and
- Completion Date

applies, as the case may be, to either the whole of the *works* or any *section* of the *works*.

OPTION X6: BONUS FOR EARLY COMPLETION

Bonus for early Completion

X6

X6.1 The *Contractor* is paid a bonus calculated at the rate stated in the Contract Data for each day from the earlier of

- Completion and
- the date on which the *Client* takes over the *works*

until the Completion Date.

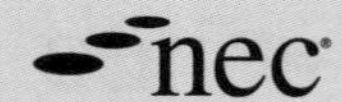

Option X7: Delay damages

Delay damages

X7

X7.1 The *Contractor* pays delay damages at the rate stated in the Contract Data from the Completion Date for each day until the earlier of

- Completion and
- the date on which the *Employer* takes over the *works*.

X7.2 If the Completion Date is changed to a later date after delay damages have been paid, the *Employer* repays the overpayment of damages with interest. Interest is assessed from the date of payment to the date of repayment and the date of repayment is an assessment date.

X7.3 If the *Employer* takes over a part of the *works* before Completion, the delay damages are reduced from the date on which the part is taken over. The *Project Manager* assesses the benefit to the *Employer* of taking over the part of the *works* as a proportion of the benefit to the *Employer* of taking over the whole of the *works* not previously taken over. The delay damages are reduced in this proportion.

OPTION X7: DELAY DAMAGES

Delay damages X7

X7.1 The *Contractor* pays delay damages at the rate stated in the Contract Data from the Completion Date for each day until the earlier of

- Completion and
- the date on which the *Client* takes over the *works*.

X7.2 If the Completion Date is changed to a later date after delay damages have been paid, the *Client* repays the overpayment of damages with interest. Interest is assessed from the date of payment to the date of repayment.

X7.3 If the *Client* takes over a part of the *works* before Completion, the delay damages are reduced from the date on which the part is taken over. The *Project Manager* assesses the benefit to the *Client* of taking over the part of the *works* as a proportion of the benefit to the *Client* of taking over the whole of the *works* not previously taken over. The delay damages are reduced in this proportion.

OPTION X8: UNDERTAKINGS TO THE *CLIENT* OR OTHERS

Undertakings to the *Client* or Others X8

X8.1 The *Contractor* gives *undertakings to Others* as stated in the Contract Data.

X8.2 If the *Contractor* subcontracts the work stated in the Contract Data it arranges for the Subcontractor to provide a *Subcontractor undertaking to Others* if required by the *Client*.

X8.3 If the *Contractor* subcontracts the work stated in the Contract Data it arranges for the Subcontractor to provide a *Subcontractor undertaking to the Client*.

X8.4 The *undertakings to Others*, *Subcontractor undertaking to Others* and *Subcontractor undertaking to the Client* are in the form set out in the Scope.

X8.5 The *Client* prepares the undertakings and sends them to the *Contractor* for signature. The *Contractor* signs the undertakings, or arranges for the Subcontractor to sign them, and returns them to the *Client* within three weeks.

The Option X8 principle has been imported from the NEC3 Professional Services Contract, but here refers to the giving of undertakings rather than the entry into collateral warranty agreements. Undertakings could include those from a Subcontractor.

OPTION X9: TRANSFER OF RIGHTS

Transfer of rights X9

X9.1 The *Client* owns the *Contractor's* rights over material prepared for the design of the *works* except as stated otherwise in the Scope. The *Contractor* obtains other rights for the *Client* as stated in the Scope and obtains from a Subcontractor equivalent rights for the *Client* over the material prepared by the Subcontractor. The *Contractor* provides to the *Client* the documents which transfer these rights to the *Client*.

Option X9 has also been imported from the NEC3 Professional Services Contract and might be used in circumstances where ownership of the material is important to the *Client*.

OPTION X10: INFORMATION MODELLING

Defined terms X10

X10.1 (1) The Information Execution Plan is the *information execution plan* or is the latest Information Execution Plan accepted by the *Project Manager*. The latest Information Execution Plan accepted by the *Project Manager* supersedes the previous Information Execution Plan.

(2) Project Information is information provided by the *Contractor* which is used to create the Information Model.

(3) The Information Model is the electronic integration of Project Information and similar information provided by the *Client* and other Information Providers and is in the form stated in the Information Model Requirements.

(4) The Information Model Requirements are the requirements identified in the Scope for creating or changing the Information Model.

(5) Information Providers are the people or organisations who contribute to the Information Model and are identified in the Information Model Requirements.

Collaboration X10.2 The *Contractor* collaborates with other Information Providers as stated in the Information Model Requirements.

Early warning X10.3 The *Contractor* and the *Project Manager* give an early warning by notifying the other as soon as either becomes aware of any matter which could adversely affect the creation or use of the Information Model.

Information Execution Plan X10.4 (1) If an Information Execution Plan is not identified in the Contract Data, the *Contractor* submits a first Information Execution Plan to the *Project Manager* for acceptance within the period stated in the Contract Data.

(2) Within two weeks of the *Contractor* submitting an Information Execution Plan for acceptance, the *Project Manager* notifies the *Contractor* of the acceptance of the Information Execution Plan or the reasons for not accepting it. A reason for not accepting an Information Execution Plan is that

- it does not comply with the Information Model Requirements or
- it does not allow the *Contractor* to Provide the Works.

If the *Project Manager* does not notify acceptance or non-acceptance within the time allowed, the *Contractor* may notify the *Project Manager* of that failure. If the failure continues for a further one week after the *Contractor's* notification, it is treated as acceptance by the *Project Manager* of the Information Execution Plan.

(3) The *Contractor* submits a revised Information Execution Plan to the *Project Manager* for acceptance

- within the *period for reply* after the *Project Manager* has instructed it to and
- when the *Contractor* chooses to.

(4) The *Contractor* provides the Project Information in the form stated in the Information Model Requirements and in accordance with the accepted Information Execution Plan.

Compensation events X10.5 If the Information Execution Plan is altered by a compensation event, the *Contractor* includes the alterations to the Information Execution Plan in the quotation for the compensation event.

Option X12: Partnering

Identified and defined terms

X12

X12.1 (1) The Partners are those named in the Schedule of Partners. The *Client* is a Partner.

(2) An Own Contract is a contract between two Partners which includes this Option.

(3) The Core Group comprises the Partners listed in the Schedule of Core Group Members.

(4) Partnering Information is information which specifies how the Partners work together and is either in the documents which the Contract Data states it is in or in an instruction given in accordance with this contract.

(5) A Key Performance Indicator is an aspect of performance for which a target is stated in the Schedule of Partners.

Actions

X12.2 (1) Each Partner works with the other Partners to achieve the *Client's objective* stated in the Contract Data and the objectives of every other Partner stated in the Schedule of Partners.

(2) Each Partner nominates a representative to act for it in dealings with other Partners.

(3) The Core Group acts and takes decisions on behalf of the Partners on those matters stated in the Partnering Information.

(4) The Partners select the members of the Core Group. The Core Group decides how they will work and decides the dates when each member joins and leaves the Core Group. The *Client's* representative leads the Core Group unless stated otherwise in the Partnering Information.

(5) The Core Group keeps the Schedule of Core Group Members and the Schedule of Partners up to date and issues copies of them to the Partners each time either is revised.

(6) This Option does not create a legal partnership between Partners who are not one of the Parties in this contract.

Use of the Information Model X10.6 The *Client* owns the Information Model and the *Contractor's* rights over Project Information except as stated otherwise in the Information Model Requirements. The *Contractor* obtains from a Subcontractor equivalent rights for the *Client* over information prepared by the Subcontractor. The *Contractor* provides to the *Client* the documents which transfer these rights to the *Client*.

Liability X10.7 (1) The following are *Client's* liabilities.

- A fault or error in the Information Model and any Project Information once it has been incorporated into the Information Model
- A fault in information provided by Information Providers other than the *Contractor*.

(2) The *Contractor* is not liable for a fault or error in the Project Information unless it failed to provide the Project Information using the skill and care normally used by professionals providing information similar to the Project Information.

(3) The *Contractor* provides insurance for claims made against it arising out of its failure to provide the Project Information using the skill and care normally used by professionals providing information similar to the Project Information. The minimum amount of this insurance is as stated in the Contract Data. This insurance provides cover from the *starting date* until the end of the period stated in the Contract Data.

Option X10 is a re-drafted version of a previous BIM clause (published in *NEC3: How to use BIM with NEC3 Contracts*), based on the CIC's BIM Protocol. The drafting now is in a neutral form, allowing an Information Execution Plan to arise at or shortly after contract award.

OPTION X11: TERMINATION BY THE *CLIENT*

Termination by the *Client* X11

X11.1 The *Client* may terminate the *Contractor's* obligation to Provide the Works for a reason not identified in the Termination Table by notifying the *Project Manager* and the *Contractor*.

X11.2 If the *Client* terminates for a reason not identified in the Termination Table the termination procedures followed are P1 and P2 and the amounts due on termination are A1, A2 and A4.

This provision was strictly available in ECC3, through the 'any other reason route', but ECC4 now follows the same approach as provided in the NEC3 Professional Services Contract by including it as a separate secondary Option.

OPTION X12: MULTIPARTY COLLABORATION (NOT USED WITH X20)

Identified and defined terms X12

X12.1 (1) Partners are those who have a contract in connection with the subject matter of the contract which includes this multiparty collaboration Option or equivalent. The *Promoter* is a Partner.

(2) The Schedule of Partners is a list of the Partners which is in the document the Contract Data states it is in and Partners subsequently added by agreement of the Partners. It sets out the objectives of the Partners and includes targets for performance.

(3) An Own Contract is a contract between two Partners.

(4) The Core Group comprises the Partners selected to take decisions on behalf of the Partners.

(5) The Schedule of Core Group Members is a list of the Partners forming the Core Group.

(6) Partnering Information is information which specifies how the Partners collaborate and is either in the documents which the Contract Data states it is in or in an instruction given in accordance with the contract.

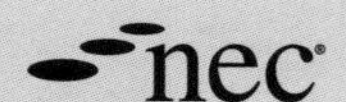

Working together

X12.3 (1) The Partners work together as stated in the Partnering Information and in a spirit of mutual trust and co-operation.

(2) A Partner may ask another Partner to provide information which he needs to carry out the work in his Own Contract and the other Partner provides it.

(3) Each Partner gives an early warning to the other Partners when he becomes aware of any matter that could affect the achievement of another Partner's objectives stated in the Schedule of Partners.

(4) The Partners use common information systems as set out in the Partnering Information.

(5) A Partner implements a decision of the Core Group by issuing instructions in accordance with its Own Contracts.

(6) The Core Group may give an instruction to the Partners to change the Partnering Information. Each such change to the Partnering Information is a compensation event which may lead to reduced Prices.

(7) The Core Group prepares and maintains a timetable showing the proposed timing of the contributions of the Partners. The Core Group issues a copy of the timetable to the Partners each time it is revised. The *Contractor* changes his programme if it is necessary to do so in order to comply with the revised timetable. Each such change is a compensation event which may lead to reduced Prices.

(8) A Partner gives advice, information and opinion to the Core Group and to other Partners when asked to do so by the Core Group. This advice, information and opinion relates to work that another Partner is to carry out under its Own Contract and is given fully, openly and objectively. The Partners show contingency and risk allowances in information about costs, prices and timing for future work.

(9) A Partner notifies the Core Group before subcontracting any work.

Incentives

X12.4 (1) A Partner is paid the amount stated in the Schedule of Partners if the target stated for a Key Performance Indicator is improved upon or achieved. Payment of the amount is due when the target has been improved upon or achieved and is made as part of the amount due in the Partner's Own Contract.

(2) The *Client* may add a Key Performance Indicator and associated payment to the Schedule of Partners but may not delete or reduce a payment stated in the Schedule of Partners.

(7) A Key Performance Indicator is an aspect of performance for which a target is stated in the Schedule of Partners.

Actions X12.2 (1) The Partners collaborate with each other to achieve the *Promoter's objective* stated in the Contract Data and the objectives of every other Partner stated in the Schedule of Partners.

(2) Each Partner nominates a representative to act for it in dealings with other Partners.

(3) The Core Group acts and takes decisions on behalf of the Partners on those matters stated in the Partnering Information.

(4) The Partners select the members of the Core Group. The Core Group decides how they will work and decides the dates when each member joins and leaves the Core Group. The *Promoter's* representative leads the Core Group unless stated otherwise in the Partnering Information.

(5) The Core Group keeps the Schedule of Core Group Members and the Schedule of Partners up to date and issues copies of them to the Partners each time either is revised.

(6) This Option does not create a legal partnership between Partners who are not one of the Parties in the contract.

Collaboration X12.3 (1) The Partners collaborate as stated in the Partnering Information and in a spirit of mutual trust and co-operation.

(2) A Partner may ask another Partner to provide information which it needs to carry out the work in its Own Contract and the other Partner provides it.

(3) Each Partner gives an early warning to the other Partners when it becomes aware of any matter that could affect the achievement of another Partner's objectives stated in the Schedule of Partners.

(4) The Partners use common information systems as set out in the Partnering Information.

(5) A Partner implements a decision of the Core Group by issuing instructions in accordance with its Own Contracts.

(6) The Core Group may give an instruction to the Partners to change the Partnering Information. Each such change to the Partnering Information is a compensation event which may lead to reduced Prices.

(7) The Core Group prepares and maintains a timetable showing the proposed timing of the contributions of the Partners. The Core Group issues a copy of the timetable to the Partners each time it is revised. The *Contractor* changes its programme if it is necessary to do so in order to comply with the revised timetable. Each such change is a compensation event which may lead to reduced Prices.

(8) A Partner gives advice, information and opinion to the Core Group and to other Partners when asked to do so by the Core Group. This advice, information and opinion relates to work that another Partner is to carry out under its Own Contract and is given fully, openly and objectively. The Partners show contingency and risk allowances in information about costs, prices and timing for future work.

(9) A Partner informs the Core Group before subcontracting any work.

Incentives X12.4 (1) A Partner is paid the amount stated in the Schedule of Partners if the target stated for a Key Performance Indicator is improved upon or achieved. Payment of the amount is due when the target has been improved upon or achieved and is made as part of the amount due in the Partner's Own Contract.

(2) The *Promoter* may add a Key Performance Indicator and associated payment to the Schedule of Partners but may not delete or reduce a payment stated in the Schedule of Partners.

Among the slight amendments made to Option X12, the important one is the change of wording from 'partnering' to 'collaboration'. As a result of ECC4's use of the word '*Client*' instead of '*Employer*', the word '*Promoter*' is used in this secondary Option. 

CORE CLAUSES
MAIN OPTION CLAUSES
SECONDARY OPTION CLAUSES
COST COMPONENTS

Option X13: Performance bond

Performance bond

X13

X13.1 The *Contractor* gives the *Employer* a performance bond, provided by a bank or insurer which the *Project Manager* has accepted, for the amount stated in the Contract Data and in the form set out in the Works Information. A reason for not accepting the bank or insurer is that its commercial position is not strong enough to carry the bond. If the bond was not given by the Contract Date, it is given to the *Employer* within four weeks of the Contract Date.

Option X14: Advanced payment to the *Contractor*

Advanced payment

X14

X14.1 The *Employer* makes an advanced payment to the *Contractor* of the amount stated in the Contract Data.

X14.2 The advanced payment is made either within four weeks of the Contract Date or, if an advanced payment bond is required, within four weeks of the later of

- the Contract Date and
- the date when the *Employer* receives the advanced payment bond.

The advanced payment bond is issued by a bank or insurer which the *Project Manager* has accepted. A reason for not accepting the proposed bank or insurer is that its commercial position is not strong enough to carry the bond. The bond is for the amount of the advanced payment which the *Contractor* has not repaid and is in the form set out in the Works Information. Delay in making the advanced payment is a compensation event.

X14.3 The advanced payment is repaid to the *Employer* by the *Contractor* in instalments of the amount stated in the Contract Data. An instalment is included in each amount due assessed after the period stated in the Contract Data has passed until the advanced payment has been repaid.

Option X15: Limitation of the *Contractor*'s liability for his design to reasonable skill and care

The *Contractor*'s design

X15

X15.1 The *Contractor* is not liable for Defects in the *works* due to his design so far as he proves that he used reasonable skill and care to ensure that his design complied with the Works Information.

X15.2 If the *Contractor* corrects a Defect for which he is not liable under this contract it is a compensation event.

OPTION X13: PERFORMANCE BOND

Performance bond X13

X13.1 The *Contractor* gives the *Client* a performance bond, provided by a bank or insurer which the *Project Manager* has accepted, for the amount stated in the Contract Data and in the form set out in the Scope. A reason for not accepting the bank or insurer is that its commercial position is not strong enough to carry the bond. If the bond was not given by the Contract Date, it is given to the *Client* within four weeks of the Contract Date.

OPTION X14: ADVANCED PAYMENT TO THE *CONTRACTOR*

Advanced payment X14

X14.1 The *Client* makes an advanced payment to the *Contractor* of the amount stated in the Contract Data. The advanced payment is included in the assessment made at the first assessment date or, if an advanced payment bond is required, at the next assessment date after the *Client* receives the advanced payment bond.

X14.2 The advanced payment bond is issued by a bank or insurer which the *Project Manager* has accepted. A reason for not accepting the proposed bank or insurer is that its commercial position is not strong enough to carry the bond. The bond is for the amount of the advanced payment which the *Contractor* has not repaid and is in the form set out in the Scope. Delay in making the advanced payment in accordance with the contract is a compensation event.

X14.3 The advanced payment is repaid to the *Client* by the *Contractor* in instalments of the amount stated in the Contract Data. An instalment is included in each amount due assessed after the period stated in the Contract Data has passed until the advanced payment has been repaid.

OPTION X15: THE *CONTRACTOR'S* DESIGN

The *Contractor's* design X15

X15.1 The *Contractor* is not liable for a Defect which arose from its design unless it failed to carry out that design using the skill and care normally used by professionals designing works similar to the works.

X15.2 If the *Contractor* corrects a Defect for which it is not liable under the contract it is a compensation event.

X15.3 The *Contractor* may use the material provided by it under the contract for other work unless

- the ownership of the material has been given to the *Client* or
- it is stated otherwise in the Scope.

X15.4 The *Contractor* retains copies of drawings, specifications, reports and other documents which record the *Contractor's* design for the *period for retention*. The copies are retained in the form stated in the Scope.

X15.5 The *Contractor* provides insurance for claims made against it arising out of its failure to use the skill and care normally used by professionals designing works similar to the *works*. The minimum amount of this insurance is as stated in the Contract Data. This insurance provides cover from the *starting date* until the end of the period stated in the Contract Data.

Option X15 is revamped to include provisions for professional indemnity insurance, document retention and future use of material, but the most important change is the reversal of the burden of proof in X15.1. 

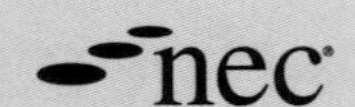

Option X16: Retention (not used with Option F)

Retention

X16

X16.1 After the Price for Work Done to Date has reached the *retention free amount*, an amount is retained in each amount due. Until the earlier of

- Completion of the whole of the *works* and
- the date on which the *Employer* takes over the whole of the *works*

the amount retained is the *retention percentage* applied to the excess of the Price for Work Done to Date above the *retention free amount*.

X16.2 The amount retained is halved

- in the assessment made at Completion of the whole of the *works* or
- in the next assessment after the *Employer* has taken over the whole of the *works* if this is before Completion of the whole of the *works*.

The amount retained remains at this amount until the Defects Certificate is issued. No amount is retained in the assessments made after the Defects Certificate has been issued.

Option X17: Low performance damages

Low performance damages

X17

X17.1 If a Defect included in the Defects Certificate shows low performance with respect to a performance level stated in the Contract Data, the *Contractor* pays the amount of low performance damages stated in the Contract Data.

OPTION X16: RETENTION (NOT USED WITH OPTION F)

Retention

X16

X16.1 After the Price for Work Done to Date has reached the *retention free amount*, an amount is retained in each amount due. Until the earlier of

- Completion of the whole of the *works* and
- the date on which the *Client* takes over the whole of the *works*

the amount retained is the *retention percentage* applied to the excess of the Price for Work Done to Date above the *retention free amount*.

X16.2 The amount retained is halved

- in the next assessment made after Completion of the whole of the *works* or
- in the next assessment after the *Client* has taken over the whole of the *works* if this is before Completion of the whole of the *works*.

The amount retained remains at this amount until the date when the Defects Certificate is due to be issued. No amount is retained in the assessments made after the Defects Certificate is due to be issued.

X16.3 If stated in the Contract Data or agreed by the *Client*, the *Contractor* may give the *Client* a retention bond, provided by a bank or insurer which the *Project Manager* has accepted, for the total amount to be retained and in the form set out in the Scope. A reason for not accepting the bank or insurer is that its commercial position is not strong enough to carry the bond. Any amount retained after the *Contractor* gives the *Client* a retention bond is paid to the *Contractor* in the next assessment.

Option X16.1 allows for a retention bond to be provided in lieu of retention, reflecting the preference of some clients. 

OPTION X17: LOW PERFORMANCE DAMAGES

Low performance damages

X17

X17.1 If a Defect included in the Defects Certificate shows low performance with respect to a performance level stated in the Contract Data, the *Contractor* pays the amount of low performance damages stated in the Contract Data.

Option X18: Limitation of liability

Limitation of liability

X18

X18.1 The *Contractor*'s liability to the *Employer* for the *Employer*'s indirect or consequential loss is limited to the amount stated in the Contract Data.

X18.2 For any one event, the liability of the *Contractor* to the *Employer* for loss of or damage to the *Employer*'s property is limited to the amount stated in the Contract Data.

X18.3 The *Contractor*'s liability to the *Employer* for Defects due to his design which are not listed on the Defects Certificate is limited to the amount stated in the Contract Data.

X18.4 The *Contractor*'s total liability to the *Employer* for all matters arising under or in connection with this contract, other than the excluded matters, is limited to the amount stated in the Contract Data and applies in contract, tort or delict and otherwise to the extent allowed under the *law of the contract.*

The excluded matters are amounts payable by the *Contractor* as stated in this contract for

- loss of or damage to the *Employer*'s property,
- delay damages if Option X7 applies,
- low performance damages if Option X17 applies and
- *Contractor*'s share if Option C or Option D applies.

X18.5 The *Contractor* is not liable to the *Employer* for a matter unless it is notified to the *Contractor* before the *end of liability date*.

Option X20: Key Performance Indicators (not used with Option X12)

Incentives

X20.1 A Key Performance Indicator is an aspect of performance by the *Contractor* for which a target is stated in the Incentive Schedule. The Incentive Schedule is the *incentive schedule* unless later changed in accordance with this contract.

X20.2 From the *starting date* until the Defects Certificate has been issued, the *Contractor* reports to the *Project Manager* his performance against each of the Key Performance Indicators. Reports are provided at the intervals stated in the Contract Data and include the forecast final measurement against each indicator.

X20.3 If the *Contractor's* forecast final measurement against a Key Performance Indicator will not achieve the target stated in the Incentive Schedule, he submits to the *Project Manager* his proposals for improving performance.

OPTION X18: LIMITATION OF LIABILITY

Limitation of liability

X18

X18.1 Each of the limits to the *Contractor's* liability in this clause apply if a limit is stated in the Contract Data.

X18.2 The *Contractor's* liability to the *Client* for the *Client's* indirect or consequential loss is limited to the amount stated in the Contract Data.

X18.3 For any one event, the liability of the *Contractor* to the *Client* for loss of or damage to the *Client's* property is limited to the amount stated in the Contract Data.

X18.4 The *Contractor's* liability to the *Client* for Defects due to its design which are not listed on the Defects Certificate is limited to the amount stated in the Contract Data.

X18.5 The *Contractor's* total liability to the *Client* for all matters arising under or in connection with the contract, other than the excluded matters, is limited to the amount stated in the Contract Data and applies in contract, tort or delict and otherwise to the extent allowed under the *law of the contract.*

The excluded matters are amounts payable by the *Contractor* as stated in the contract for

- loss of or damage to the *Client's* property,
- delay damages if Option X7 applies,
- low performance damages if Option X17 applies and
- *Contractor's* share if Option C or Option D applies.

X18.6 The *Contractor* is not liable to the *Client* for a matter unless details of the matter are notified to the *Contractor* before the *end of liability date.*

OPTION X20: KEY PERFORMANCE INDICATORS (NOT USED WITH OPTION X12)

Incentives

X20

X20.1 A Key Performance Indicator is an aspect of performance by the *Contractor* for which a target is stated in the Incentive Schedule. The Incentive Schedule is the *incentive schedule* unless later changed in accordance with the contract.

X20.2 From the *starting date* until the Defects Certificate has been issued, the *Contractor* reports to the *Project Manager* its performance against each of the Key Performance Indicators. Reports are provided at the intervals stated in the Contract Data and include the forecast final measurement against each indicator.

X20.3 If the *Contractor's* forecast final measurement against a Key Performance Indicator will not achieve the target stated in the Incentive Schedule, it submits to the *Project Manager* its proposals for improving performance.

X20.4 The *Contractor* is paid the amount stated in the Incentive Schedule if the target stated for a Key Performance Indicator is improved upon or achieved. Payment of the amount is due when the target has been improved upon or achieved.

X20.5 The *Employer* may add a Key Performance Indicator and associated payment to the Incentive Schedule but may not delete or reduce a payment stated in the Incentive Schedule.

X20.4 The *Contractor* is paid the amount stated in the Incentive Schedule if the target stated for a Key Performance Indicator is improved upon or achieved. Payment of the amount is due when the target has been improved upon or achieved.

X20.5 The *Client* may add a Key Performance Indicator and associated payment to the Incentive Schedule but may not delete or reduce a payment stated in the Incentive Schedule.

OPTION X21: WHOLE LIFE COST

Whole life cost **X21**

X21.1 The *Contractor* may propose to the *Project Manager* that the Scope is changed in order to reduce the cost of operating and maintaining an asset.

X21.2 If the *Project Manager* is prepared to consider the change, the *Contractor* submits a quotation which comprises

- a detailed description,
- the forecast cost reduction to the *Client* of the asset over its whole life,
- an analysis of the resulting risks to the *Client*,
- the proposed changes to the Prices and
- a revised programme showing any changes to the Completion Date and Key Dates.

X21.3 The *Project Manager* consults with the *Contractor* about a quotation. The *Project Manager* replies within the *period for reply*. The reply is acceptance of the quotation or the reasons for not accepting it. The *Project Manager* may give any reason for not accepting the quotation.

X21.4 The *Project Manager* does not change the Scope as proposed by the *Contractor* unless the *Contractor's* quotation is accepted.

X21.5 When a quotation to reduce the costs of operating and maintaining an asset is accepted the *Project Manager* changes the Scope, the Prices, the Completion Date and the Key Dates accordingly and accepts the revised programme. The change to the Scope is not a compensation event.

Option X21 is a new secondary Option designed to reduce the cost of operating and maintaining an asset by encouraging the *Contractor* to suggest changes to the Scope. 

OPTION X22: EARLY *CONTRACTOR* INVOLVEMENT (USED ONLY WITH OPTIONS C AND E)

Identified and defined terms **X22**

X22.1 (1) Budget is the items and amounts stated in the Contract Data unless the amounts are later changed in accordance with the contract.

(2) Project Cost is the total paid by the *Client* to the *Contractor* and Others for the items included in the Budget.

(3) Stage One and Stage Two have the meanings given to them in the Scope.

(4) Pricing Information is information which specifies how the *Contractor* prepares its assessment of the Prices for Stage Two, and is in the document which the Contract Data states it is in.

Forecasts X22.2 (1) The *Contractor* provides detailed forecasts of the total Defined Cost of the work to be done in Stage One for acceptance by the *Project Manager*. Forecasts are prepared at the intervals stated in the Contract Data from the *starting date* until the issue of a notice to proceed to Stage Two.

(2) Within one week of the *Contractor* submitting a forecast for acceptance, the *Project Manager* either accepts the forecast or notifies the *Contractor* of the reasons for not accepting it. A reason for not accepting the forecast is that

- it does not comply with the Scope or
- it includes work which is not necessary for Stage One.

(3) The *Contractor* makes a revised submission taking account of the *Project Manager's* reasons.

(4) The cost of any work that is not included in the accepted forecast is treated as a Disallowed Cost.

(5) The *Contractor* prepares forecasts of the Project Cost in consultation with the *Project Manager* and submits them to the *Project Manager*. Forecasts are prepared at the intervals stated in the Contract Data from the *starting date* until Completion of the whole of the *works*. An explanation of the changes made since the previous forecast is submitted with each forecast.

Proposals for Stage Two X22.3

(1) The *Contractor* submits its design proposals for Stage Two to the *Project Manager* for acceptance in accordance with the submission procedure stated in the Scope.

(2) The submission includes the *Contractor's* forecast of the effect of the design proposal on the Project Cost and the Accepted Programme.

(3) If the submission is not accepted, the *Project Manager* gives reasons. A reason for not accepting a *Contractor's* submission is that

- it does not comply with the Scope,
- it will cause the *Client* to incur unnecessary costs to Others or
- the *Project Manager* is not satisfied that the Prices or any changes to the Prices have been properly assessed.

(4) The *Contractor* makes a revised submission taking account of the *Project Manager's* reasons.

(5) The total of the Prices for Stage Two is assessed by the *Contractor* using the Pricing Information stated in the Contract Data.

[Option C only] X22.4 (1) If the main Option is C the *Contractor* submits the total of the Prices for Stage Two to the *Project Manager* in the form of revisions to the Activity Schedule. The Activity Schedule includes the Price for Work Done to Date in Stage One.

(2) The *Contractor* obtains approvals and consents from Others as stated in the Scope.

(3) Any additional Scope provided by the *Contractor* in Stage One becomes Scope provided by the *Contractor* for its design.

(4) The *Contractor* completes any outstanding design during Stage Two.

Key persons X22.5 The *Contractor* does not replace any *key person* during Stage One unless

- the *Project Manager* instructs the *Contractor* to do so or
- the person is unable to continue to act in connection with the contract.

Notice to proceed to Stage Two X22.6

(1) The *Project Manager* issues a notice to proceed to Stage Two when

- the *Contractor* has obtained approvals and consents from Others as stated in the Scope,
- changes to the Budget have been agreed or assessed by the *Project Manager*,
- the *Project Manager* and the *Contractor* have agreed the total of the Prices for Stage Two and
- the *Client* has confirmed the *works* are to proceed.

(2) If a notice to proceed to Stage Two is not issued for any reason, the *Project Manager* issues an instruction that the work required in Stage Two is removed from the Scope. This instruction is not a compensation event.

(3) If the *Project Manager* does not issue a notice to proceed to Stage Two because

- the *Project Manager* and the *Contractor* have not agreed the total of the Prices for Stage Two or
- the *Contractor* has failed to achieve the performance requirements stated in the Scope the *Client* may appoint another contractor to complete the Stage Two *works*.

Changes to the Budget X22.7

(1) If one of the following events happens, the *Project Manager* and the *Contractor* discuss different ways of dealing with changes to the Budget which are practicable.

- The *Project Manager* gives an instruction changing the *Client's* requirements stated in the Scope.
- Additional events stated in the Contract Data.

(2) The *Project Manager* and the *Contractor* agree changes to the Budget within four weeks of the event arising which changes the Budget. If the *Project Manager* and the *Contractor* cannot agree the changes to the Budget the *Project Manager* assesses the change and notifies the *Contractor* of the assessment.

Incentive payment X22.8

(1) If the final Project Cost is less than the Budget, the *Contractor* is paid the *budget incentive*. The *budget incentive* is calculated by multiplying the difference between the Budget and the final Project Cost by the percentage stated in the Contract Data.

(2) The *Project Manager* makes a preliminary assessment of the *budget incentive* at Completion of the whole of the *works* and includes this in the amount due following Completion of the whole of the *works*.

(3) The *Project Manager* makes a final assessment of the *budget incentive* and includes this in the final amount due.

Option X22 is a new secondary Option, but it was previously available as an Option Z clause at www.neccontract.com. 

CORE CLAUSES

MAIN OPTION CLAUSES

SECONDARY OPTION CLAUSES

COST COMPONENTS

OPTION Y

Option Y(UK)1: Project Bank Account

Definitions **Y(UK)1**

Y1.1 (1) The Authorisation is a document authorising the *project bank* to make payments to the *Contractor* and Named Suppliers.

(2) Named Suppliers are *named suppliers* and other Suppliers who have signed the Joining Deed.

(3) Project Bank Account is the account used to receive payments from the *Employer* and the *Contractor* and make payments to the *Contractor* and Named Suppliers.

(4) A Supplier is a person or organisation who has a contract to

- construct or install part of the *works*,
- provide a service necessary to Provide the Works or
- supply Plant and Materials for the *works*.

(5) Trust Deed is an agreement in the form set out in the contract which contains provisions for administering the Project Bank Account.

(6) Joining Deed is an agreement in the form set out in the contract under which the Supplier joins the Trust Deed.

Project Bank Account

Y1.2 The *Contractor* establishes the Project Bank Account with the *project bank* within three weeks of the Contract Date.

Y1.3 Unless stated otherwise in the Contract Data, the *Contractor* pays any charges made and is paid any interest paid by the *project bank*. The charges and interest by the *project bank* are not included in Defined Cost.

Y1.4 The *Contractor* submits to the *Project Manager* for acceptance details of the banking arrangements for the Project Bank Account. A reason for not accepting the banking arrangements is that they do not provide for payments to be made in accordance with this contract. The *Contractor* provides to the *Project Manager* copies of communications with the *project bank* in connection with the Project Bank Account.

Named Suppliers

Y1.5 The *Contractor* includes in his contracts with Named Suppliers the arrangements in this contract for the operation of the Project Bank Account and Trust Deed. The *Contractor* notifies the Named Suppliers of the details of the Project Bank Account and the arrangements for payment of amounts due under their contracts.

Y1.6 The *Contractor* submits proposals for adding a Supplier to the Named Suppliers to the *Project Manager* for acceptance. A reason for not accepting is that the addition of the Supplier does not comply with the Works Information. The *Employer*, the *Contractor* and the Supplier sign the Joining Deed after acceptance.

Option Y

OPTION Y(UK)1: PROJECT BANK ACCOUNT

Project Bank Account	Y(UK)1	
Definitions	Y1.1	(1) The Authorisation is a document authorising the *project bank* to make payments to the *Contractor* and Named Suppliers. (2) Named Suppliers are *named suppliers* and other Suppliers who have signed the Joining Deed. (3) Project Bank Account is the account used to receive payments from the *Client* and the *Contractor* and to make payments to the *Contractor* and Named Suppliers. (4) A Supplier is a person or organisation who has a contract to • construct or install part of the *works*, • provide a service necessary to Provide the Works or • supply Plant and Materials for the *works*. (5) Trust Deed is an agreement in the form set out in the contract which contains provisions for administering the Project Bank Account. (6) Joining Deed is an agreement in the form set out in the contract under which the Supplier joins the Trust Deed.
Project Bank Account	Y1.2	The *Contractor* establishes the Project Bank Account with the *project bank* within three weeks of the Contract Date.
	Y1.3	Unless stated otherwise in the Contract Data, the *Contractor* pays any charges made and is paid any interest paid by the *project bank*. The charges and interest by the *project bank* are not included in Defined Cost.
	Y1.4	The *Contractor* submits to the *Project Manager* for acceptance details of the banking arrangements for the Project Bank Account. A reason for not accepting the banking arrangements is that they do not provide for payments to be made in accordance with the contract. The *Contractor* provides to the *Project Manager* copies of communications with the *project bank* in connection with the Project Bank Account.
Named Suppliers	Y1.5	The *Contractor* includes in its contracts with Named Suppliers the arrangements in the contract for the operation of the Project Bank Account and Trust Deed. The *Contractor* informs the Named Suppliers of the details of the Project Bank Account and the arrangements for payment of amounts due under their contracts.
	Y1.6	The *Contractor* submits proposals for adding a Supplier to the Named Suppliers to the *Project Manager* for acceptance. A reason for not accepting is that the addition of the Supplier does not comply with the Scope. The *Client*, the *Contractor* and the Supplier sign the Joining Deed after acceptance.

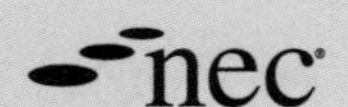

Payments

Y1.7 On or before each assessment date, the *Contractor* submits to the *Project Manager* an application for payment, and shows in the application the amounts due to Named Suppliers in accordance with their contracts.

Y1.8 Within the time set out in the banking arrangements to allow the *project bank* to make payment to the *Contractor* and Named Suppliers in accordance with the contract,

- the *Employer* makes payment to the Project Bank Account of the amount which is due to be paid under the contract and
- the *Contractor* makes payment to the Project Bank Account of any amount which the *Employer* has notified the *Contractor* he intends to withhold from the certified amount and which is required to make payment to Named Suppliers.

Y1.9 The *Contractor* prepares the Authorisation, setting out the sums due to Named Suppliers as assessed by the *Contractor* and to the *Contractor* for the balance of the payment due under the contract. After signing the Authorisation, the *Contractor* submits it to the *Project Manager* no later than four days before the final date for payment. The *Employer* signs the Authorisation and submits it to the *project bank* no later than one day before the final date for payment.

Y1.10 The *Contractor* and Named Suppliers receive payment from the Project Bank Account of the sums set out in the Authorisation as soon as practicable after the Project Bank Account receives payment.

Y1.11 A payment which is due from the *Contractor* to the *Employer* is not made through the Project Bank Account.

Effect of payment

Y1.12 Payments made from the Project Bank Account are treated as payments from the *Employer* to the *Contractor* in accordance with this contract or from the *Contractor* or *Subcontractor* to Named Suppliers in accordance with their contracts as applicable. A delay in payment due to a failure of the *Contractor* to comply with the requirements of this clause is not treated as late payment under this contract.

Trust Deed

Y1.13 The *Employer*, the *Contractor* and *named suppliers* sign the Trust Deed before the first assessment date.

Termination

Y1.14 If the *Project Manager* issues a termination certificate, no further payment is made into the Project Bank Account.

Payments	Y1.7	The *Contractor* shows in the application for payment the amounts due to Named Suppliers in accordance with their contracts.
	Y1.8	Within the time set out in the banking arrangements to allow the *project bank* to make payment to the *Contractor* and Named Suppliers in accordance with the contract, • the *Client* makes payment to the Project Bank Account of the amount which is due to be paid under the contract and • the *Contractor* makes payment to the Project Bank Account of any amount which the *Client* has informed the *Contractor* it intends to withhold from the certified amount and which is required to make payment to Named Suppliers.
	Y1.9	The *Contractor* prepares the Authorisation, setting out the sums due to Named Suppliers as assessed by the *Contractor* and to the *Contractor* for the balance of the payment due under the contract. After signing the Authorisation, the *Contractor* submits it to the *Project Manager* no later than four days before the final date for payment. The *Client* signs the Authorisation and submits it to the *project bank* no later than one day before the final date for payment.
	Y1.10	The *Contractor* and Named Suppliers receive payment from the Project Bank Account of the sums set out in the Authorisation as soon as practicable after the Project Bank Account receives payment.
	Y1.11	A payment which is due from the *Contractor* to the *Client* is not made through the Project Bank Account.
Effect of payment	Y1.12	Payments made from the Project Bank Account are treated as payments from the *Client* to the *Contractor* in accordance with the contract or from the *Contractor* or Subcontractor to Named Suppliers in accordance with their contracts as applicable. A delay in payment due to a failure of the *Contractor* to comply with the requirements of this clause is not treated as late payment under the contract.
Trust Deed	Y1.13	The *Client*, the *Contractor* and *named suppliers* sign the Trust Deed before the first assessment date.
Termination	Y1.14	If the *Project Manager* issues a termination certificate, no further payment is made into the Project Bank Account.

Trust Deed

This agreement is made between the *Employer*, the *Contractor* and the Named Suppliers.

Terms in this deed have the meanings given to them in the contract between . . . . and . . . . for . . . . (the *works*).

Background

The *Employer* and the *Contractor* have entered into a contract for the *works*.

The Named Suppliers have entered into contracts with the *Contractor* or a Subcontractor in connection with the *works*.

The *Contractor* has established a Project Bank Account to make provision for payment to the *Contractor* and the Named Suppliers.

Agreement

The parties to this deed agree that

- sums due to the *Contractor* and Named Suppliers and set out in the Authorisation are held in trust in the Project Bank Account by the *Contractor* for distribution to the *Contractor* and Named Suppliers in accordance with the banking arrangements applicable to the Project Bank Account,
- further Named Suppliers may be added as parties to this deed with the agreement of the *Employer* and *Contractor*. The agreement of the *Employer* and *Contractor* is treated as agreement by the Named Suppliers who are parties to this deed,
- this deed is subject to the law of the contract for the *works*,
- the benefits under this deed may not be assigned.

Executed as a deed on . . . .

by

. . . . (*Employer*)

. . . . (*Contractor*)

. . . .

. . . .

. . . .

. . . .

(Named Suppliers)

TRUST DEED

This agreement is made between the *Client*, the *Contractor* and the Named Suppliers.

Terms in this deed have the meanings given to them in the contract between . . . . and . . . . for . . . . (the *works*).

Background

The *Client* and the *Contractor* have entered into a contract for the *works*.

The Named Suppliers have entered into contracts with the *Contractor* or a Subcontractor in connection with the *works*.

The *Contractor* has established a Project Bank Account to make provision for payment to the *Contractor* and the Named Suppliers.

Agreement

The parties to this deed agree that

- sums due to the *Contractor* and Named Suppliers and set out in the Authorisation are held in trust in the Project Bank Account by the *Contractor* for distribution to the *Contractor* and Named Suppliers in accordance with the banking arrangements applicable to the Project Bank Account,
- further Named Suppliers may be added as parties to this deed with the agreement of the *Client* and *Contractor*. The agreement of the *Client* and *Contractor* is treated as agreement by the Named Suppliers who are parties to this deed,
- this deed is subject to the law of the contract for the *works*,
- the benefits under this deed may not be assigned.

Executed as a deed on . . . .

by

. . . . (*Client*)

. . . . (*Contractor*)

. . . .

. . . .

. . . .

. . . .

(Named Suppliers)

Joining Deed

This agreement is made between the *Employer*, the *Contractor* and . . . . (the Additional Supplier).

Terms in this deed have the meanings given to them in the contract between . . . . and . . . . for . . . . (the *works*).

Background

The *Employer* and the *Contractor* have entered into a contract for the *works*.

The Named Suppliers have entered into contracts with the *Contractor* or a Subcontractor in connection with the *works*.

The *Contractor* has established a Project Bank Account to make provision for payment to the *Contractor* and the Named Suppliers.

The *Employer*, the *Contractor* and the Named Suppliers have entered into a deed as set out in Annex 1 (the Trust Deed), and have agreed that the Additional Supplier may join that deed.

Agreement

The Parties to this deed agree that

- the Additional Supplier becomes a party to the Trust Deed from the date set out below,
- this deed is subject to the law of the contract for the *works*,
- the benefits under this deed may not be assigned.

Executed as a deed on . . . .

by

. . . . (*Employer*)

. . . . (*Contractor*)

. . . . (Additional Supplier)

JOINING DEED

This agreement is made between the *Client*, the *Contractor* and . . . . (the Additional Supplier).

Terms in this deed have the meanings given to them in the contract between . . . . and . . . . for . . . . (the *works*).

Background

The *Client* and the *Contractor* have entered into a contract for the *works*.

The Named Suppliers have entered into contracts with the *Contractor* or a Subcontractor in connection with the *works*.

The *Contractor* has established a Project Bank Account to make provision for payment to the *Contractor* and the Named Suppliers.

The *Client*, the *Contractor* and the Named Suppliers have entered into a deed as set out in Annex 1 (the Trust Deed), and have agreed that the Additional Supplier may join that deed.

Agreement

The parties to this deed agree that

- the Additional Supplier becomes a party to the Trust Deed from the date set out below,
- this deed is subject to the law of the contract for the *works*,
- the benefits under this deed may not be assigned.

Executed as a deed on . . . .

by

. . . . (*Client*)

. . . . (*Contractor*)

. . . . (Additional Supplier)

Option Y(UK)2: The Housing Grants, Construction and Regeneration Act 1996

Y(UK)2

Definitions

Y2.1 (1) The Act is the Housing Grants, Construction and Regeneration Act 1996 as amended by the Local Democracy, Economic Development and Construction Act 2009.

(2) A period of time stated in days is a period calculated in accordance with Section 116 of the Act.

Dates for payment

Y2.2 The date on which a payment becomes due is seven days after the assessment date.

The final date for payment is fourteen days or a different period for payment if stated in the Contract Data after the date on which payment becomes due.

The *Project Manager*'s certificate is the notice of payment to the *Contractor* specifying the amount due at the payment due date (the notified sum) and stating the basis on which the amount was calculated.

Notice of intention to pay less

Y2.3 If either Party intends to pay less than the notified sum, he notifies the other Party not later than seven days (the prescribed period) before the final date for payment by stating the amount considered to be due and the basis on which that sum is calculated. A Party does not withhold payment of an amount due under this contract unless he has notified his intention to pay less than the notified sum as required by this contract.

Suspension of performance

Y2.4 If the *Contractor* exercises his right under the Act to suspend performance, it is a compensation event.

OPTION Y(UK)2: THE HOUSING GRANTS, CONSTRUCTION AND REGENERATION ACT 1996

The Housing Grants, Construction and Regeneration Act 1996 Y(UK)2

Definitions

Y2.1 In this Option, time periods stated in days exclude Christmas Day, Good Friday and bank holidays.

Dates for payment

Y2.2 The date on which a payment becomes due is seven days after the assessment date. The date on which the final payment becomes due is

- if the *Project Manager* makes an assessment after the issue of a Defects Certificate, five weeks after the issue of the Defects Certificate,
- if the *Project Manager* does not make an assessment after the issue of a Defects Certificate, one week after the *Contractor* issues its assessment or
- if the *Project Manager* has issued a termination certificate, fourteen weeks after the issue of the certificate.

The final date for payment is fourteen days after the date on which payment becomes due or a different period for payment if stated in the Contract Data.

The *Project Manager's* certificate is the notice of payment specifying the amount due at the payment due date (the notified sum, which may be zero) and stating the basis on which the amount was calculated. If the *Project Manager* does not make an assessment after the issue of a Defects Certificate, the *Contractor's* assessment is the notice of payment.

Notice of intention to pay less

Y2.3 If either Party intends to pay less than the notified sum, it notifies the other Party not later than seven days (the prescribed period) before the final date for payment by stating the amount considered to be due and the basis on which that sum is calculated. A Party does not withhold payment of an amount due under the contract unless it has notified its intention to pay less than the notified sum as required by the contract.

Y2.4 If the *Client* terminates for one of reasons R1 to R15, R18 or R22 and a certified payment has not been made at the date of the termination certificate, the *Client* makes the certified payment unless

- it has notified the *Contractor* in accordance with the contract that it intends to pay less than the notified sum or
- the termination is for one of reasons R1 to R10 and the reason occurred after the last date on which it could have notified the *Contractor* in accordance with the contract that it intends to pay less than the notified sum.

Suspension of performance

Y2.5 If the *Contractor* exercises its right under the Housing Grants, Construction and Regeneration Act 1996 as amended by the Local Democracy, Economic Development and Construction Act 2009 to suspend performance, it is a compensation event.

> Some additional provisions are included in Y(UK)2 to define the date on which a final payment becomes due in certain circumstances and what happens in the event that the *Client* terminates for one of the reasons listed.
>
> 

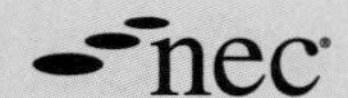

Option Y(UK)3: The Contracts (Rights of Third Parties) Act 1999

Third party rights

Y(UK)3

Y3.1 A person or organisation who is not one of the Parties may enforce a term of this contract under the Contracts (Rights of Third Parties) Act 1999 only if the term and the person or organisation are stated in the Contract Data.

Option Z: *Additional conditions of contract*

Additional conditions of contract

Z1

Z1.1 The *additional conditions of contract* stated in the Contract Data are part of this contract.

OPTION Y(UK)3: THE CONTRACTS (RIGHTS OF THIRD PARTIES) ACT 1999

Third party rights

Y(UK)3

Y3.1 A *beneficiary* may enforce the terms of the contract stated in the Contract Data under the Contracts (Rights of Third Parties) Act 1999.

Y3.2 Other than the Parties or a *beneficiary*, no person can enforce any of the terms of the contract under the Contracts (Rights of Third Parties) Act 1999.

Y3.3 If a *beneficiary* is identified by class or description and not as a named person or organisation, the *Client* notifies the *Contractor* of the name of the *beneficiary* once they have been identified.

OPTION Z: *ADDITIONAL CONDITIONS OF CONTRACT*

Additional conditions of contract

Z1

Z1.1 The *additional conditions of contract* stated in the Contract Data are part of the contract.

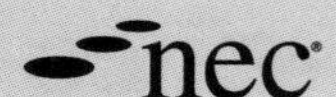

GLOSSARY OF NEW ECC4 TERMS	
beneficiary	new identified term for Option Y(UK)3 – a third party rights beneficiary
Budget	new defined term in Option X22 – the early contractor involvement provisions
budget incentive	new identified term for Option X22 – the *Contractor's* proportion of any savings that arise due to early contractor involvement
Client	changed project role – formerly '*Employer*'
Collaboration	new contract term – replaces previous phrase 'working together' in title of X12.3, Option X12
Early *Contractor* involvement	title of new Option X22
Information Execution Plan	defined term in new Option X10 – the plan defining protocols and procedures for the supply and use of Project Information
information execution plan	identified term for new Option X10 – the named document containing the Information Execution Plan
Information Model	defined term in new Option X10 – an electronic model integrating Project Information and other data as specified in the Scope
Information Model Requirements	defined term in new Option X10 – requirements in the Scope for how the Information Model is to be created and developed
Information modelling	title of new Option X10 – Option covering the implementation of information within an NEC project
Information Providers	defined Term in new Option X10 – people or organisations who provide information for the Information Model
Multiparty collaboration	new title for Option X12 – formerly 'Partnering'
PAF	new abbreviation used in Option X1 – for 'Price Adjustment Factor'
period for retention	new identified term for Option X15 – the period for retention of documents recording the *Contractor's* design
Pricing Information	new defined term in Option X22 – specification of how the *Contractor* prepares its prices for Stage Two of the early contractor involvement procedure
Project Cost	new defined term in Option X22 – the total amount paid by the *Client*
Project Information	defined term in new Option X10 – information provided by the *Contractor* for the Information Model
Promoter	new contract role – replaces '*Client*' in Option X12
Promoter's objective	changed identified term – replaces '*Client's objective*' in Option X12
retention bond	new contract term in Option X16 – a bond provided by the *Contractor* to the *Client* for the amount of any retention
Stage One and Stage Two	new defined term in Option X22 – defining the early contractor involvement procedure, as specified in the Scope
Subcontractor undertakings to Others	identified term for new Option X8
Subcontractor undertakings to the Client	identified term for new Option X8
Termination by the *Client*	title of new Option X11
Transfer of rights	title of new Option X9 – covering rights over material prepared by the *Contractor* and *Subcontractor*
ultimate holding company	changed contract term in Option X4 – formerly 'parent company'
undertakings to Others	new identified term for Option X8 – undertakings by the *Contractor*
Undertakings to the *Client* or Others	title of new Option X8
Whole life cost	title of new Option X21 – new option to incentivise reduction of costs in use

SCHEDULE OF COST COMPONENTS

This schedule is part of the *conditions of contract* only when Option C, D or E is used. In this schedule the *Contractor* means the *Contractor* and not his Subcontractors. An amount is included only in one cost component and only if it is incurred in order to Provide the Works.

People

1 The following components of the cost of

- people who are directly employed by the *Contractor* and whose normal place of working is within the Working Areas and
- people who are directly employed by the *Contractor* and whose normal place of working is not within the Working Areas but who are working in the Working Areas.

11 Wages, salaries and amounts paid by the *Contractor* for people paid according to the time worked while they are within the Working Areas.

12 Payments to people for

(a) bonuses and incentives

(b) overtime

(c) working in special circumstances

(d) special allowances

(e) absence due to sickness and holidays

(f) severance related to work on this contract.

13 Payments made in relation to people for

(a) travel

(b) subsistence and lodging

(c) relocation

(d) medical examinations

(e) passports and visas

(f) travel insurance

(g) items (a) to (f) for dependants

(h) protective clothing

(i) meeting the requirements of the law

(j) pensions and life assurance

Schedule of Cost Components

This schedule is part of these *conditions of contract* only when Option C, D or E is used. An amount is included

- only in one cost component, and
- only if it is incurred in order to Provide the Works.

People

1 The following components of

- the cost of people who are directly employed by the *Contractor* and whose normal place of working is within the Working Areas and
- the cost of people who are directly employed by the *Contractor* and whose normal place of working is not within the Working Areas but who are working in the Working Areas, proportionate to the time they spend working in the Working Areas.

The addition to the second bullet confirms what users would have usually interpreted the previous provision to mean. 

11 Wages, salaries and amounts paid by the *Contractor* for people paid according to the time worked on the contract.

The clarification to item 11 again reflects the position that most users would have taken in the previous edition. 

12 Payments related to work on the contract and made to people for

(a) bonuses and incentives

(b) overtime

(c) working in special circumstances

(d) special allowances

(e) absence due to sickness and holidays

(f) severance.

13 Payments made in relation to people in accordance with their employment contract for

(a) travel

(b) subsistence and lodging

(c) relocation

(d) medical examinations

(e) passports and visas

(f) travel insurance

(g) items (a) to (f) for dependants

(h) protective clothing

(i) contributions, levies or taxes imposed by law

(j) pensions and life assurance

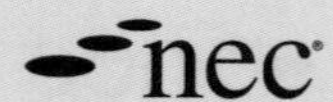

(k) death benefit

(l) occupational accident benefits

(m) medical aid

(n) a vehicle

(o) safety training.

14 The following components of the cost of people who are not directly employed by the *Contractor* but are paid for by him according to the time worked while they are within the Working Areas.

Amounts paid by the *Contractor*.

Equipment

2 The following components of the cost of Equipment which is used within the Working Areas (including the cost of accommodation but excluding Equipment cost covered by the percentage for Working Areas overheads).

21 Payments for the hire or rent of Equipment not owned by

- the *Contractor*,
- his parent company or
- by a company with the same parent company

at the hire or rental rate multiplied by the time for which the Equipment is required.

22 Payments for Equipment which is not listed in the Contract Data but is

- owned by the *Contractor*,
- purchased by the *Contractor* under a hire purchase or lease agreement or
- hired by the *Contractor* from the *Contractor*'s parent company or from a company with the same parent company

at open market rates, multiplied by the time for which the Equipment is required.

23 Payments for Equipment purchased for work included in this contract listed with a time-related on cost charge, in the Contract Data, of

- the change in value over the period for which the Equipment is required and
- the time-related on cost charge stated in the Contract Data for the period for which the Equipment is required.

The change in value is the difference between the purchase price and either the sale price or the open market sale price at the end of the period for which the Equipment is required. Interim payments of the change in value are made at each assessment date. A final payment is made in the next assessment after the change in value has been determined.

If the *Project Manager* agrees, an additional item of Equipment may be assessed as if it had been listed in the Contract Data.

24 Payments for special Equipment listed in the Contract Data. These amounts are the rates stated in the Contract Data multiplied by the time for which the Equipment is required.

If the *Project Manager* agrees, an additional item of special Equipment may be assessed as if it had been listed in the Contract Data.

(k) death benefit

(l) occupational accident benefits

(m) medical aid and health insurance

(n) a vehicle

(o) safety training.

The addition in item 13 can be evidenced using employment contracts where necessary. 

14 The following components of the cost of people who are not directly employed by the *Contractor* but are paid for by the *Contractor* according to the time worked while they are within the Working Areas.

Amounts paid by the *Contractor.*

Equipment

2 The following components of the cost of Equipment which is used within the Working Areas.

21 Payments for the hire or rent of Equipment not owned by

- the *Contractor,*
- the *Contractor's* ultimate holding company or
- a company with the same ultimate holding company

at the hire or rental rate multiplied by the time for which the Equipment is required.

22 Payments for Equipment which is not listed in the Contract Data but is

- owned by the *Contractor,*
- purchased by the *Contractor* under a hire purchase or lease agreement or
- hired by the *Contractor* from the *Contractor's* ultimate holding company or from a company with the same ultimate holding company

at open market rates, multiplied by the time for which the Equipment is required.

23 Payments for Equipment purchased for work included in the contract listed with a time-related on cost charge, in the Contract Data, of

- the change in value over the period for which the Equipment is required and
- the time-related on cost charge stated in the Contract Data for the period for which the Equipment is required.

The change in value is the difference between the purchase price and either the sale price or the open market sale price at the end of the period for which the Equipment is required. Interim payments of the change in value are made at each assessment date. A final payment is made in the next assessment after the change in value has been determined.

If the *Project Manager* agrees, an additional item of Equipment may be assessed as if it had been listed in the Contract Data.

24 Payments for special Equipment listed in the Contract Data. These amounts are the rates stated in the Contract Data multiplied by the time for which the Equipment is required.

If the *Project Manager* agrees, an additional item of special Equipment may be assessed as if it had been listed in the Contract Data.

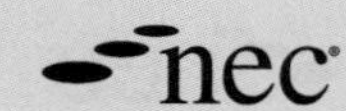

25 Payments for the purchase price of Equipment which is consumed.

26 Unless included in the hire or rental rates, payments for

- transporting Equipment to and from the Working Areas other than for repair and maintenance,
- erecting and dismantling Equipment and
- constructing, fabricating or modifying Equipment as a result of a compensation event.

27 Payments for purchase of materials used to construct or fabricate Equipment.

28 Unless included in the hire rates, the cost of operatives is included in the cost of people.

Plant and Materials

3 The following components of the cost of Plant and Materials.

31 Payments for

- purchasing Plant and Materials,
- delivery to and removal from the Working Areas,
- providing and removing packaging and
- samples and tests.

32 Cost is credited with payments received for disposal of Plant and Materials unless the cost is disallowed.

Charges

4 The following components of the cost of charges paid by the *Contractor.*

41 Payments for provision and use in the Working Areas of

- water,
- gas and
- electricity.

42 Payments to public authorities and other properly constituted authorities of charges which they are authorised to make in respect of the *works*.

43 Payments for

(a) cancellation charges arising from a compensation event

(b) buying or leasing land

(c) compensation for loss of crops or buildings

(d) royalties

(e) inspection certificates

(f) charges for access to the Working Areas

(g) facilities for visits to the Working Areas by Others

(h) specialist services

(i) consumables and equipment provided by the *Contractor* for the *Project Manager*'s and *Supervisor*'s offices.

25 Payments for the purchase price of Equipment which is consumed.

26 Unless included in the hire or rental rates, payments for

- transporting Equipment to and from the Working Areas other than for repair and maintenance,
- erecting and dismantling Equipment and
- constructing, fabricating or modifying Equipment as a result of a compensation event.

27 Payments for purchase of materials used to construct or fabricate Equipment.

28 Unless included in the hire rates, the cost of operatives is included in the cost of people.

Plant and Materials

3 The following components of the cost of Plant and Materials.

31 Payments for

- purchasing Plant and Materials,
- delivery to and removal from the Working Areas,
- providing and removing packaging and
- samples and tests.

32 Cost is credited with payments received for disposal of Plant and Materials unless the cost is disallowed.

Subcontractors

4 The following components of the cost of Subcontractors.

The provisions for dealing with the cost of Subcontractors has been moved from the definition of Defined Cost into the rules laid down in the Schedule of Cost Components. 

41 Payments to Subcontractors for work which is subcontracted without taking into account any amounts paid to or retained from the Subcontractor by the *Contractor*, which would result in the *Client* paying or retaining the amount twice.

Charges

5 The following components of the cost of charges paid or received by the *Contractor*.

51 Payments for the provision and use in the Working Areas of

- water,
- gas,
- electricity,
- telephone and
- internet.

52 Payments to public authorities and other properly constituted authorities of charges which they are authorised to make in respect of the *works*.

53 Payments for

(a) cancellation charges arising from a compensation event

(b) buying or leasing land or buildings within the Working Area

(c) compensation for loss of crops or buildings

(d) royalties

(e) inspection certificates

(f) charges for access to the Working Areas

(g) facilities for visits to the Working Areas by Others

(h) consumables and equipment provided by the *Contractor* for the *Project Manager's* and *Supervisor's* offices.

44 A charge for overhead costs incurred within the Working Areas calculated by applying the percentage for Working Areas overheads stated in the Contract Data to the total of people items 11, 12, 13 and 14. The charge includes provision and use of equipment, supplies and services, but excludes accommodation, for

(a) catering

(b) medical facilities and first aid

(c) recreation

(d) sanitation

(e) security

(f) copying

(g) telephone, telex, fax, radio and CCTV

(h) surveying and setting out

(i) computing

(j) hand tools not powered by compressed air.

Manufacture and fabrication

5 The following components of the cost of manufacture and fabrication of Plant and Materials which are

- wholly or partly designed specifically for the *works* and
- manufactured or fabricated outside the Working Areas.

51 The total of the hours worked by employees multiplied by the hourly rates stated in the Contract Data for the categories of employees listed.

52 An amount for overheads calculated by multiplying this total by the percentage for manufacturing and fabrication overheads stated in the Contract Data.

Design

6 The following components of the cost of design of the *works* and Equipment done outside the Working Areas.

61 The total of the hours worked by employees multiplied by the hourly rates stated in the Contract Data for the categories of employees listed.

62 An amount for overheads calculated by multiplying this total by the percentage for design overheads stated in the Contract Data.

63 The cost of travel to and from the Working Areas for the categories of design employees listed in the Contract Data.

Insurance

7 The following are deducted from cost

- the cost of events for which this contract requires the *Contractor* to insure and
- other costs paid to the *Contractor* by insurers.

54 Payments made and received by the *Contractor* for the removal from Site and disposal or sale of materials from excavation and demolition.

The most radical change to the rules in 'Charges' is the removal of item 44 and the related percentage for Working Areas overheads. The matters previously covered are now distributed across appropriate components in the Schedule. 

Manufacture and fabrication

6 The following components of the cost of manufacture and fabrication of Plant and Materials which are

- wholly or partly designed specifically for the *works* and
- manufactured or fabricated by the *Contractor* outside the Working Areas.

61 The total of the hours worked by people multiplied by the hourly rates stated in the Contract Data for the categories of people listed.

Design

7 The following components of the cost of design of the *works* and Equipment done outside the Working Areas.

71 The total of the hours worked by people multiplied by the hourly rates stated in the Contract Data for the categories of people listed.

In both the 'Manufacture and fabrication' and 'Design' components, the provisions for overheads have been deleted to simplify matters. Such overheads now need to be allowed for in the associated hourly rates in the Contract Data part two. 

72 The cost of travel to and from the Working Areas for the categories of design employees listed in the Contract Data.

Insurance

8 The following are deducted from cost

- the cost of events for which the contract requires the *Contractor* to insure and
- other costs paid to the *Contractor* by insurers.

SHORTER SCHEDULE OF COST COMPONENTS

This schedule is part of the *conditions of contract* only when Option A, B, C, D or E is used. When Option C, D or E is used, this schedule is used by agreement for assessing compensation events. When Option C, D or E is used, in this schedule the *Contractor* means the *Contractor* and not his Subcontractors. An amount is included only in one cost component and only if it is incurred in order to Provide the Works.

People

1 The following components of the cost of

- people who are directly employed by the *Contractor* and whose normal place of working is within the Working Areas,
- people who are directly employed by the *Contractor* and whose normal place of working is not within the Working Areas but who are working in the Working Areas and
- people who are not directly employed by the *Contractor* but are paid for by him according to the time worked while they are within the Working Areas.

11 Amounts paid by the *Contractor* including those for meeting the requirements of the law and for pension provision.

Equipment

2 The following components of the cost of Equipment which is used within the Working Areas (including the cost of accommodation but excluding Equipment cost covered by the percentage for people overheads).

21 Amounts for Equipment which is in the published list stated in the Contract Data. These amounts are calculated by applying the percentage adjustment for listed Equipment stated in the Contract Data to the rates in the published list and by multiplying the resulting rate by the time for which the Equipment is required.

22 Amounts for Equipment listed in the Contract Data which is not in the published list stated in the Contract Data. These amounts are the rates stated in the Contract Data multiplied by the time for which the Equipment is required.

23 The time required is expressed in hours, days, weeks or months consistently with the list of items of Equipment in the Contract Data or with the published list stated in the Contract Data.

24 Unless the item is in the published list and the rate includes the cost component, payments for

- transporting Equipment to and from the Working Areas other than for repair and maintenance,
- erecting and dismantling Equipment and
- constructing, fabricating or modifying Equipment as a result of a compensation event.

25 Unless the item is in the published list and the rate includes the cost component, the purchase price of Equipment which is consumed.

26 Unless included in the rate in the published list, the cost of operatives is included in the cost of people.

Short Schedule of Cost Components

Note the minor change in the Schedule's title, from 'Shorter' to 'Short'. This Schedule can now only be used with Option A or B to assess compensation events.

This schedule is part of these *conditions of contract* only when Option A or B is used. An amount is included

- only in one cost component, and
- only if it is incurred in order to Provide the Works.

People

1 The following components of the cost of

- people who are directly employed by the *Contractor* and whose normal place of working is within the Working Areas,
- people who are directly employed by the *Contractor* and whose normal place of working is not within the Working Areas but who are working in the Working Areas, proportionate to the time they spend working in the Working Areas and
- people who are not directly employed by the *Contractor* but are paid for by it according to the time worked while they are within the Working Areas.

11 Amounts calculated by multiplying each of the People Rates by the total time appropriate to that rate spent within the Working Areas.

This Schedule adopts a pre-pricing approach to people, using the *people rates* in Contract Data part two, rather than requiring their Defined Cost to be determined, as provided for in the previous edition.

Equipment

2 The following components of the cost of Equipment which is used within the Working Areas.

21 Amounts for Equipment which is in the published list stated in the Contract Data. These amounts are calculated by applying the percentage adjustment for listed Equipment stated in the Contract Data to the rates in the published list and by multiplying the resulting rate by the time for which the Equipment is required.

22 Amounts for Equipment listed in the Contract Data which is not in the published list stated in the Contract Data. These amounts are the rates stated in the Contract Data multiplied by the time for which the Equipment is required.

23 The time required is expressed in hours, days, weeks or months consistently with the list of items of Equipment in the Contract Data or with the published list stated in the Contract Data.

24 Unless the item is in the published list and the rate includes the cost component, payments for

- transporting Equipment to and from the Working Areas other than for repair and maintenance,
- erecting and dismantling Equipment and
- constructing, fabricating or modifying Equipment as a result of a compensation event.

25 Unless the item is in the published list and the rate includes the cost component, the purchase price of Equipment which is consumed.

26 Unless included in the rate in the published list, the cost of operatives is included in the cost of people.

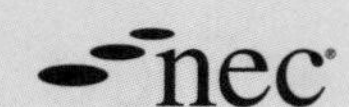

27 Amounts for Equipment which is neither in the published list stated in the Contract Data nor listed in the Contract Data, at competitively tendered or open market rates, multiplied by the time for which the Equipment is required.

Plant and Materials

3 The following components of the cost of Plant and Materials.

31 Payments for

- purchasing Plant and Materials,
- delivery to and removal from the Working Areas,
- providing and removing packaging and
- samples and tests.

32 Cost is credited with payments received for disposal of Plant and Materials unless the cost is disallowed.

Charges

4 The following components of the cost of charges paid by the *Contractor.*

41 A charge calculated by applying the percentage for people overheads stated in the Contract Data to people item 11 to cover the costs of

- payments for the provision and use in the Working Areas of water, gas and electricity,
- payments for buying or leasing land, compensation for loss of crops or buildings, royalties, inspection certificates, charges for access to the Working Areas, facilities for visits to the Working Areas by Others and
- payments for the provision and use of equipment,supplies and services (excluding accommodation) for catering, medical facilities and first aid, recreation, sanitation, security, copying, telephones, telex, fax, radio, CCTV, surveying, setting out, computing and hand tools not powered by compressed air.

42 Payments for cancellation charges arising from a compensation event.

43 Payments to public authorities and other properly constituted authorities of charges which they are authorised to make in respect of the *works.*

44 Consumables and equipment provided by the *Contractor* for the *Project Manager*'s and *Supervisor*'s office.

45 Specialist services.

27 Amounts for Equipment which is neither in the published list stated in the Contract Data nor listed in the Contract Data, at competitively tendered or open market rates, multiplied by the time for which the Equipment is required.

Plant and Materials

3 The following components of the cost of Plant and Materials.

31 Payments for

- purchasing Plant and Materials,
- delivery to and removal from the Working Areas,
- providing and removing packaging and
- samples and tests.

32 Cost is credited with payments received for disposal of Plant and Materials unless the cost is disallowed.

Subcontractors

4 The following components of the cost of Subcontractors.

41 Payments to Subcontractors for work which is subcontracted.

Another significant change to the Short Schedule is that Subcontractors are recognised as a cost component, unlike in the previous edition. This helps to simplify assessment of compensation events, along with reverting to the single Fee.

Charges

5 The following components of the cost of charges paid or received by the *Contractor.*

51 Payments for the provision and use in the Working Areas of

- water,
- gas,
- electricity,
- telephone and
- internet.

52 Payments to public authorities and other properly constituted authorities of charges which they are authorised to make in respect of the *works*.

53 Payments for

(a) cancellation charges arising from a compensation event

(b) buying or leasing land or buildings within the Working Area

(c) compensation for loss of crops or buildings

(d) royalties

(e) inspection certificates

(f) charges for access to the Working Areas

(g) facilities for visits to the Working Areas by Others

(h) consumables and equipment provided by the *Contractor* for the *Project Manager's* and *Supervisor's* offices.

54 Payments made and received by the *Contractor* for the removal from Site and disposal or sale of materials from excavation and demolition.

Considerable changes are made to 'Charges', including the removal of the percentage for people overheads. The provisions now line up with those of the Schedule of Cost Components. The changes should also assist users by standardising and simplifying this Schedule.

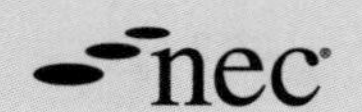

Manufacture and fabrication

5 The following components of the cost of manufacture and fabrication of Plant and Materials, which are

- wholly or partly designed specifically for the *works* and
- manufactured or fabricated outside the Working Areas.

51 Amounts paid by the *Contractor*.

Design

6 The following components of the cost of design of the *works* and Equipment done outside the Working Areas.

61 The total of the hours worked by employees multiplied by the hourly rates stated in the Contract Data for the categories of employees listed.

62 An amount for overheads calculated by multiplying this total by the percentage for design overheads stated in the Contract Data.

63 The cost of travel to and from the Working Areas for the categories of design employees listed in the Contract Data.

Insurance

7 The following are deducted from cost

- costs against which this contract required the *Contractor* to insure and
- other costs paid to the *Contractor* by insurers.

Manufacture and fabrication

6 The following components of the cost of manufacture and fabrication of Plant and Materials, which are

- wholly or partly designed specifically for the *works* and
- manufactured or fabricated by the *Contractor* outside the Working Areas.

61 The total of the hours worked by people multiplied by the hourly rates stated in the Contract Data for the categories of people listed.

The 'Manufacture and fabrication' provisions now align with those in the Schedule of Cost Components. 

Design

7 The following components of the cost of design of the *works* and Equipment done outside the Working Areas.

71 The total of the hours worked by employees multiplied by the hourly rates stated in the Contract Data for the categories of employees listed.

In the changed 'Design' provisions, the overheads component has been deleted to simplify matters. Such overheads now need to be allowed for in the associated hourly rates in the Contract Data part two. 

72 The cost of travel to and from the Working Areas for the categories of design employees listed in the Contract Data.

Insurance

8 The following are deducted from cost

- the cost of events for which the contract requires the *Contractor* to insure and
- other costs paid to the *Contractor* by insurers.